Noel D. Uri, AB, MA, MS, PhD

Conservation Tillage in U.S. Agriculture
Environmental, Economic, and Policy Issues

Pre-publication
REVIEWS,
COMMENTARIES,
EVALUATIONS . . .

"The author provides comprehensive insights into the physical, historical, economic, social, and policy aspects of soil erosion and conservation tillage. In this book the likely benefits and costs associated with the use of conservation tillage that would accrue to farmers and to the public are thoroughly identified and quantified.

It is an excellent source of information and data on all aspects of conservation tillage that students, researchers, farmers, and policy makers will find highly useful. I found the material provided in the sections on the private and social benefits and cost of conservation tillage especially valuable."

Kazim Konyar, PhD
Associate Professor,
California State University,
San Bernardino

Food Products Press
An Imprint of The Haworth Press, Inc.

Conservation Tillage in U.S. Agriculture

Environmental, Economic, and Policy Issues

THE FOOD PRODUCTS PRESS
Crop Science
Amarjit S. Basra, PhD
Senior Editor

New, Recent, and Forthcoming Titles of Related Interest:

Dictionary of Plant Genetics and Molecular Biology by Gurbachan S. Miglani

Advances in Hemp Research by Paolo Ranalli

Wheat: Ecology and Physiology of Yield Determination by Emilio H. Satorre and Gustavo A. Slafer

Mineral Nutrition of Crops: Fundamental Mechanisms and Implications by Zdenko Rengel

Conservation Tillage in U.S. Agriculture: Environmental, Economic, and Policy Issues by Noel D. Uri

Cotton Fibers: Developmental Biology, Quality Improvement, and Textile Processing by Amarjit S. Basra

Intensive Cropping: Efficient Use of Water, Nutrients, and Tillage by S. S. Prihar, P. R. Gajri, D. K. Benbi, and V. K. Arora

Plant Growth Regulators in Agriculture and Horticulture: Role and Commerical Uses edited by Amarjit S. Basra

Crop Responses and Adaptations to Temperature Stress: New Insights and Approaches edited by Amarjit S. Basra

Physiological Bases for Maize Improvement edited by Maria Elena Otegui and Gustavo A. Slafer

Conservation Tillage in U.S. Agriculture
Environmental, Economic, and Policy Issues

Noel D. Uri, AB, MA, MS, PhD

Food Products Press
An Imprint of The Haworth Press, Inc.
New York • London • Oxford

Published by

Food Products Press, an imprint of The Haworth Press, Inc., 10 Alice Street, Binghamton, NY 13904-1580

Softcover edition published 2000.

Cover design by Monica L. Seifert.

The Library of Congress has cataloged the hardcover edition of this book as:

Uri, Noel D.
    Conservation tillage in U.S. agriculture : environmental, economic, and policy issues / Noel D. Uri.
        p. cm.
    Includes bibliographical references (p.   ) and index.
    ISBN 1-56022-884-9 (alk. paper)
    1. Conservation tillage—United States. I. Title. II. Title: Conservation tillage in U.S. agriculture. III. Title: Conservation tillage in United States agriculture.
S604.U75 1999
338.1′62—dc21   98-43756
                                                                                  CIP

ISBN 1-56022-897-0 (pbk.)

CONTENTS

ABOUT THE AUTHOR

Noel D. Uri, PhD, is currently with the Resource Inventory Division of the Natural Resources Conservation Service at the U.S. Department of Agriculture in Washington, DC. Dr. Uri has held a number of federal government and academic positions. Previously, he worked for the Bureau of Labor Statistics, the Federal Energy Administration, the U.S. Department of Energy, the Federal Trade Commission, and the U.S. Department of Agriculture. He has also been on the faculties of the Catholic University of America and George Mason University. Dr. Uri has published several hundred journal articles on agricultural, environmental, and resource issues.

Foreword

Soil erosion associated with agricultural production practices can impose significant private and social costs. Severe soil degradation from erosion can destroy the productive capacity of the soil. It can also impair water quality through sediment and agricultural chemicals. Three related causes of water quality impairment are sedimentation, eutrophication, and pesticide contamination. When soil particles and agricultural chemicals wash off a field, they may be carried in runoff until discharged into a water body or stream. Not all agricultural constituents that are transported from a field reach water systems, but a significant portion does, especially dissolved chemicals and the more chemically active, finer soil particles. Once agricultural pollutants enter a water system, they lower water quality and can impose economic losses on water users. These off-site impacts can be substantial. The off-site impacts of erosion are potentially greater than the on-site productivity effects in the aggregate. Therefore, society may have a stronger incentive for reducing erosion than agricultural producers.

Conservation tillage is one of many conservation practices developed to reduce soil erosion. Conservation tillage practices that leave substantial amounts of crop residue evenly distributed over the soil surface defend against the potential of rainfall's kinetic energy to generate sediment and increase water runoff. Many field studies conducted under natural rainfall conditions on highly erodible land have compared erosion rates among tillage systems. Compared with conventional tillage practices, conservation tillage generally reduces soil erosion by 50 percent or more.

The adoption of conservation tillage has been increasing in the United States in response to growing concerns about the impact of agricultural production on the environment. An identifiable upward trend in the use of conservation tillage persisted until 1995; since then, little discernible change has occurred.

The concern over the impact of agricultural production on the environment suggests that government agencies primarily concerned with agriculture will need to enhance their capacity to address agricultural resource management issues. This need is becoming acute as other public and private entities concerned with agriculture and the environment are becoming increasingly active in this area. The need to improve policymaking analysis with respect to environmental and resource management is obvious. The problems in agriculture are shared with other sectors of the economy because agriculture is a major user of land, water, and agrichemicals, and these factors affect environmental quality, recreation, wildlife, urban water supplies, and human health. Consequently, agricultural problems are an extremely important part of overall environmental and resource issues. This book aids greatly in enhancing our understanding of these issues and in identifying available solutions.

A number of themes run throughout this book. First, agricultural producers do respond to economic factors, which can take a number of forms, including market price signals and government subsidies. Second, the government can influence the choice of production practices through a variety of means, including regulation, educational and technical assistance, research and development, taxes, and subsidies. Third, government intervention can be warranted in some instances due to the presence of externalities, such as groundwater contamination. Finally, agriculture in the United States is dynamic, continually adopting new production practices and technologies. It is through a combination of these factors that the negative impacts of agriculture on the environment can be reduced.

Wen-Yuan Huang
Economic Research Service
U.S. Department of Agriculture

Preface

During the past two decades, there has been a major shift by U.S. farmers away from conventional tillage involving inversion of the soil toward systems relying on conservation tillage that involves reduced or no tillage. The development and use of conservation tillage as a viable production practice has brought with it both private and social benefits and cost. These benefits and costs have many components, some of which are readily quantifiable, while others are more ambiguous and therefore not easily measured. This study assesses both the private and the social benefits and costs associated with the adoption of conservation tillage.

Conservation tillage has played a central role in agricultural program policy in the United States. For most of the sixty-plus-year history of federal conservation programs, emphasis was placed on the productivity consequences of reducing soil erosion (with benefits accruing to farmers). Beginning in the 1970s, an added concern was the increasing interest in environmental issues. Conservation programs continued to target enhancements in soil productivity and net farm income, but the focus was expanded to reduce off-farm impacts of agriculture on the environment (with benefits accruing to society as a whole).

Active federal involvement in soil conservation began in the 1930s with the authorization of a study to examine the causes of erosion and to recommend methods for its control in response to the dust storms in Oklahoma during this period. The Conservation Technical Assistance (CTA) program was established in the Soil Conservation and Domestic Allotment Act of 1935. It was designed to assist farmers in planning and installing approved conservation

---

The author's views expressed in this book do not necessarily represent the policies of the U.S. Department of Agriculture or the views of other U.S. Department of Agriculture staff members.

ix

measures to protect agricultural land from soil erosion. The CTA program was followed in 1936 by the creation of the Agricultural Conservation Program (ACP), which provided cost-sharing assistance to farmers for implementing conservation practices designed to restore and improve soil fertility and to minimize erosion caused by wind and water. The program's general approach was to share with farmers the costs of installing structures for soil building and soil conserving practices.

With the emphasis in the 1970s on expanding agricultural production in response to higher real output prices, farmers responded by moving away from long-term rotations to more continuous row cropping. Consequently, soil erosion increased and surface water quality deteriorated. In response, federal conservation policy increasingly stressed production practices to mitigate the off-farm effects of sediment and other pollutants generated by agriculture. During the 1970s and continuing on into the 1980s, the emphasis changed from implementing individual conservation practices to implementing multiple best management practices (BMPs).

Conservation efforts took on a decidedly different look in 1985. The Food Security Act of 1985 targeted highly erodible land under the conservation compliance program and the sodbuster provision. The provision stipulated that under the conservation compliance program, farmers producing crops on previously cultivated highly erodible land while participating in U.S. Department of Agriculture programs could lose eligibility for program benefits. Under conservation compliance, farmers with highly erodible cropland were required to have a conservation plan approved by January 1, 1990, and that plan had to be implemented by January 1, 1995. Many different conservation tillage practices were acceptable.

The most recent manifestation of agricultural program policy is the Federal Agriculture Improvement and Reform (FAIR) Act of 1996. It modifies the conservation compliance provisions of the Food Security Act of 1985 to provide farmers with greater flexibility in developing and implementing conservation plans, in self-certifying compliance, and in obtaining variances for problems affecting application of conservation plans. Producers who violate conservation plans or who fail to use a conservation system on highly erodible land risk loss of eligibility for many payments, including production flexibility contract payments. An important aspect of this act is that in self-certifying compliance, there is no requirement that a status review be conducted for producers who self-certify. The FAIR Act also does not differentiate between previously cultivated and uncultivated land, thereby eliminating the sodbuster provision. Newly cropped, highly erodible land may use conservation systems other than the systems previously required under the sodbuster provision.

Conservation tillage is a viable production practice not just on highly erodible land. Many of the previously noted benefits and costs associated with conservation compliance are applicable to cropland that is not classified as highly erodible. Successful crop production is a combination of many elements, including two that are essential: the proper management of inputs and a thorough understanding of the soil resources and how they respond to production practices.

The decision by a farmer to adopt a conservation tillage practice will lead to an increase in net private benefits. However, spillover effects also occur so that, although the benefits to society as a whole of a farmer adopting conservation tillage on land that is not highly erodible will not be as great as they are on highly erodible land in a given area, these benefits are still potentially significant.

In this book, substantive insights are provided into the factors affecting the benefits and costs of the adoption of conservation tillage.

Noel D. Uri

Chapter 1

Introduction

Soil erosion from cropland in the United States has been recognized as an important problem for over sixty years. Concern initially was centered on the loss of fertile topsoil and the fear that agricultural productivity would decline. Recently, however, there has been a growing understanding of the off-farm impacts of sediment and chemical transport (U.S. Environmental Protection Agency, 1995). Therefore, both private (on-farm) and public (off-farm) benefits can be gained from reducing erosion from the nation's agricultural lands.

Conservation tillage is one of many conservation practices developed to reduce soil erosion. In its broadest sense, conservation tillage is defined as a tillage system that leaves enough crop residue on the field after harvest to protect the soil from erosion. In general, tillage that leaves a residue cover of at least 30 percent after planting is deemed conservation tillage; residue cover will vary, however, according to soil type, slope, crop rotation, winter crop cover, and other factors.

Additional benefits are associated with the use of conservation tillage beyond keeping the soil on the field. After several years under the practice, the soil's organic matter and structure may improve, thus increasing the quality of the agricultural soil. The change in organic content and the lack of soil disturbance also serve to sequester carbon, which may have long-term environmental benefits. Cropland on which conservation tillage is used also can serve as important habitat for wildlife. The residue left on the field offers food for some species and shelter for others.

Conservation tillage is not used on most U.S. cropland. The decision to change production technologies is based on many factors. Farmers will adopt conservation tillage if they perceive a gain in net benefits from switching technologies. These benefits can represent more than just the direct monetary factors reflected on a business balance sheet. Farmers also include nonmonetary adjustment costs, such as having to learn new skills or deal with new suppliers, when they assess whether to change production practices. What is not typically included in farmers' private decisions are the benefits or costs to society associated with the use of a new practice.

The purpose of this book is to identify and quantify, as much as possible, the likely benefits and costs associated with the use of conservation tillage that would accrue to farmers and to the public.

The current status of conservation tillage adoption is described in Chapter 2. The use of the technology varies widely by crop and by region, and factors affecting the adoption decision are discussed.

In Chapter 3, the on-farm and off-farm benefits and costs of conservation tillage adoption are identified. Conservation tillage and conventional tillage are compared with respect to yields and costs of production. Differences in input use are also described. As with many resource-conserving technologies, the relative advantage of conservation tillage depends on farm and regional characteristics (Caswell, Zilberman, and Casterline, 1993). The off-site or off-farm impacts of soil erosion, particularly with respect to water quality, are identified, and the benefits of tillage adoption on wildlife habitat and the reduction of carbon emissions are discussed.

An analysis is conducted to estimate the reduction of soil erosion that would result from the adoption of conservation tillage on lands for which the technology is considered suitable. Using figures developed by Ribaudo (1989) and Huzsar and Piper (1986), estimates are made of the public benefits that would be realized from the adoption-induced reductions in off-site erosion impacts. The results of this analysis are presented in Chapter 3.

If the off-site and on-site benefits of increasing the use of conservation tillage are greater than the costs to farmers of adopting the technology, several public policies can help influence farmers to adopt conservation tillage. A lexicon of these policy options is provided in Chapter 4. The U.S. government, primarily through the U.S. Department of Agriculture (USDA), has developed a suite of policies to promote the use of preferred agricultural practices.

These policies and their effect on the adoption of conservation tillage are also described.

Before the discussion of current conditions and policies, let us first consider a brief history of conservation tillage in the United States.

CONSERVATION TILLAGE IN HISTORICAL CONTEXT

The use of crop residue in the United States to sustain soil productivity dates to the early eighteenth century. Colonists from Europe, observing the more severe storms in the New World than in the Old, realized the need to use cover crops and crop residue to mitigate soil erosion (Moldenhauer, Kemper, and Langdale, 1994). Throughout the nineteenth century, farmers experimented with alternative production practices to produce food and fiber without degrading the soil resources in the process. It was only in the 1920s, however, that field experiments were initiated to assess objectively the effects of sheet and rill erosion[1] and wind erosion on soil productivity. These early efforts were disparate and tended to focus on issues of concern to a specific geographic area, for example, corn in Illinois (Odell et al., 1984). Unfortunately, no objective measures are available that describe how these efforts affected farmer behavior and net farm income (private benefits) or the extent to which soil erosion was reduced (public benefits) and soil productivity enhanced.

With the establishment of the Soil Conservation Service in 1935, a more organized and comprehensive assessment of conservation tillage began. A large number of conservation tillage practices, such as mulch tillage, were evaluated at land grant university experiment stations throughout the United States (Moldenhauer, Kemper, and Langdale, 1994). It was quickly realized that due to spatial variation in soil characteristics and weather patterns, farming with conservation tillage required a different approach to soil preparation, fertilizer application, and weed and insect control, as well as an awareness of the topography of the land farmed in relation to water sources. Therefore, conservation tillage technology was difficult to transfer unilaterally across major soil resource areas. Thus, conservation tillage diversity would be the norm, meaning that conservation tillage practices would have to be tailored to the specific crops grown and to the climatic conditions in a geographic region. A characteristic that all such practices would possess, however, was a need for the farmer to understand and use appropriate management practices.

Following World War II, plow planting methods were refined by the U.S. Department of Agriculture and land grant university researchers. It was shown that the best soil conservation contribution of these methods was surface roughness to control runoff. Although cool season crop residues were managed near the soil surface, some secondary cultivation was required for weed control, even though selective herbicides were available.

Many other forms of conservation tillage emerged during the 1960s and 1970s, including ridge tillage for cold wet soils of the Corn Belt and strip till for restrictive horizon Ultisols of the Southeast. It was during this time period that the use of conservation tillage became widespread.

The following chapters will place in perspective the role that conservation tillage can play in sustaining agricultural productivity and improving the environment.

Chapter 2

Current Status
of Conservation Tillage

- Conservation tillage was used on nearly 36 percent of planted acreage in 1996. This level has remained relatively unchanged over the past few years.
- The use of conservation tillage varies by crop and is dependent on site-specific factors, including soil type, topsoil depth, and local climatic conditions.
- A number of economic, demographic, geographic, and policy factors have affected the adoption of conservation tillage. It is not possible to quantify exactly the impact of these factors on conservation tillage use.
- Management complexities and profitability are key factors impeding the adoption of conservation tillage.

The adoption of conservation tillage and the current extent of its use depend on a variety of economic, demographic, geographic, and policy factors. This chapter describes the evolution of the use of conservation tillage and provides some insights into the importance of economic and environmental considerations in a farmer's decision to adopt its practices.

DEFINITION OF CONSERVATION TILLAGE

Conservation tillage has evolved from tillage practices that range from reducing the number of trips over the field to raising crops without primary or secondary tillage. Emphasis today is on leaving crop residues on the soil surface after planting rather than merely reducing the number of trips across the field, although these two practices are closely related.

As early as 1963, the Soil Conservation Service of the U.S. Department of Agriculture began tracking the number of cropland acres planted by minimum tillage, as it was referred to at the time. In that year, minimum tillage was reportedly used on about 3.8 million planted acres or about 1 percent of the total number of acres. By 1967, the number of minimum tillage acres had doubled (Mannering, Schertz, and Julian, 1987).

One of the difficulties in tracking the trend in conservation tillage use has been the absence of any consistent definition of conservation tillage until relatively recently. Before 1977, minimum tillage was used to describe a conservation tillage practice primarily aimed at reducing the number of tillage trips over a field (Langdale et al., 1992). A large portion of the acres on which minimum tillage was used would have had considerable amounts of residue after planting, but a significant portion also would not have met the current definition of conservation tillage (Schertz, 1988).

In late 1977, the Soil Conservation Service changed the term minimum tillage to conservation tillage and defined it as a form of noninversion tillage that retains protective amounts of residue mulch on the surface throughout the year. Types of conservation tillage included no tillage, strip tillage, stubble mulching, and other types of noninversion tillage.

In early 1984, the Soil Conservation Service changed the definition of conservation tillage once again.[1] This definition remains the one commonly used (Conservation Technology Information Center [CTIC], 1996, p. 6):

Any tillage and planting system that maintains at least 30 percent of the soil surface covered by residue after planting to reduce soil erosion by water. Where soil erosion by wind is the primary concern, any system that maintains at least 1,000 pounds (per acre) of flat, small grain residue equivalent on the surface during the critical wind erosion period. Two key factors influencing crop residue are (1) the type of crop, which establishes the initial residue amount and determines its fragility, and (2) the type of tillage operations prior to and including planting.[2] (See Figure 2.1.)

FIGURE 2.1. Tillage System Definitions

Crop Residue Management (CRM)—A year-round conservation system that usually involves a reduction in the number of passes over the field with tillage implements and/or in the intensity of tillage operations, including the elimination of plowing (inversion of the surface layer of soil). CRM begins with the selection of crops that produce sufficient quantities of residue to reduce wind and water erosion and may include the use of cover crops after low residue-producing crops. CRM includes all field operations that affect residue amounts, orientation, and distribution throughout the period requiring protection. The amounts of residue cover needed at specific sites are usually expressed in percentage but may also be in pounds. Tillage systems included under CRM are conservation tillage (no tillage, ridge tillage, and mulch tillage) and reduced tillage.

Conservation Tillage—Any tillage and planting system that maintains at least 30 percent of the soil surface covered by residue after planting to reduce soil erosion by water. Where soil erosion by wind is the primary concern, any system that maintains at least 1,000 pounds (per acre) of flat, small grain equivalent on the surface during the critical wind erosion period. Two key factors influencing crop residue are (1) the type of crop, which establishes the initial residue amount and determines its fragility, and (2) the type of tillage operations prior to and including planting.

Conservation Tillage Systems include:

Mulch tillage—The soil is disturbed prior to planting. Tillage tools such as chisels, field cultivators, disks, sweeps, or blades are used. Weed control is accomplished with herbicides and/or cultivation.

Ridge tillage—The soil is left undisturbed from harvest to planting except for nutrient injection. Planting is completed in a seedbed prepared on ridges with sweeps, disk openers, coulters, or row cleaners. Residue is left on the surface between ridges. Weed control is accomplished with herbicides and/or cultivation. Ridges are rebuilt during cultivation.

No tillage—The soil is left undisturbed from harvest to planting except for nutrient injection. Planting or drilling is accomplished in a narrow seedbed or slot created by coulters, row cleaners, disk openers, in-row chisels, or rototillers. Weed control is accomplished primarily with herbicides. Cultivation may be used for emergency weed control.

Reduced-till (15 to 30 percent residue)—Tillage types that leave 15 to 30 percent residue cover after planting, or 500 to 1,000 pounds per acre of small grain residue equivalent, throughout the critical wind erosion period. Weed control is accomplished with herbicides and/or cultivation.

Conventional-till (less than 15 percent residue)—Tillage types that leave less than 15 percent residue cover after planting, or less than 500 pounds per acre of small grain residue equivalent, throughout the critical wind erosion period. Generally includes plowing or other intensive tillage. Weed control is accomplished with herbicides and/or cultivation.

Source: Conservation Technology Information Center (1996, p. 6).

CONSERVATION TILLAGE DATA

Sources of information on the use of conservation tillage are somewhat limited. Currently, two primary sources exist and each of these is relied upon in what follows. Because each is somewhat different, it is important to understand how the data are compiled. The first source is the Conservation Technology Information Center's *National Crop Residue Management Survey.* In 1983, the Conservation Tillage Information Center (CTIC)[3] was established, its primary objective being to serve as a clearinghouse for gathering and sharing information on conservation tillage. The CTIC is a division of the National Association of Conservation Districts and is supported by agribusiness, government (federal and state), and other organizations. The definition of conservation tillage adopted by the Soil Conservation Service was developed in cooperation with the CTIC. The CTIC conducts an annual survey of crop residue management practices to provide accurate acreage and residue data to gauge the use of crop residue management systems among farmers in the United States. The survey, which includes all planting and tillage types, is conducted on a county-by-county basis that yields approximately 3,100 responses. The enumerators are a panel of local directors of U.S. Department of Agriculture program agencies (Natural Resources Conservation Service [NRCS], formerly the Soil Conservation Service, Cooperative Extension Service, and Farm Service Agency, formerly the Agricultural Stabilization and Conservation Service) and others knowledgeable about local crop residue management practices. These enumerators complete the survey each summer. The survey provides only information about crop residue management systems. It does not provide information on crop yields, production costs, input use, whether the cropland is classified as highly erodible, what other production practices are used (e.g., crop rotations), or the costs of implementing any of the conservation tillage practices.

The second source of information about the use of conservation tillage is the *Cropping Practices Survey* (CPS) conducted annually by the U.S. Department of Agriculture between 1990 and 1995. Annual data were collected on fertilizer and pesticide use, tillage systems employed, cropping sequences, whether the cropland is designated as highly erodible, and the use of other inputs and production practices.

The survey covered corn, cotton, soybeans, wheat (winter, spring, and durum), and potatoes. Only selected states were surveyed, but about 80 percent of the total planted acreage for the respective crops is covered. Five tillage systems, including conventional tillage with moldboard plow, conventional tillage without moldboard plow, mulch tillage, ridge tillage, and no tillage, are defined based on the use of specific tillage implements and their residue incorporation rates (Bull, 1993). The CPS was not designed to collect information on whether fields meet the conservation compliance requirements nor to reveal the costs of implementing conservation tillage practices.[4] Although the CPS did collect information on tillage and planting implements used, it did not collect comprehensive machinery use information such as the type of tractor or its horsepower rating. No survey collects this latter information.

TRENDS IN CONSERVATION TILLAGE USE

The use of conservation tillage in the United States experienced an identifiable upward trend until the last few years, which show no discernible change. A longer-term perspective can be obtained from Figure 2.2. The use of conservation tillage increased from 2 percent of planted acreage in 1968 to nearly 36 percent of planted acreage in 1996 (Schertz, 1988; CTIC, 1996).[5] A disaggregated view of the use of conservation tillage can be obtained by looking at specific crops and states/regions. Because the CTIC data are more comprehensive than the CPS data, CTIC data will be used to summarize conservation tillage adoption levels. CTIC data collection at the crop/state level based on a consistent definition of conservation tillage began only in 1989. Hence, the data reported start in that year.

The percentage of U.S. cropland that was conservation tilled increased from 26 percent (71.7 million acres) to 36 percent (103.8 million acres) over the seven-year period (1989 to 1996), but the increase differs by crop (see Table 2.1a and Table 2.1b). That is, conservation tillage is used mostly on soybeans, corn, and small grains. More than 40 percent of total corn and soybean planted acreage in 1996 was conservation tilled. Where double-cropping was used, nearly 70 percent of soybean acreage, 46 percent of corn acreage, and 37 percent of sorghum acreage employed conservation tillage systems.

FIGURE 2.2. Percent of Planted Acres on Which Conservation Tillage Is Adopted

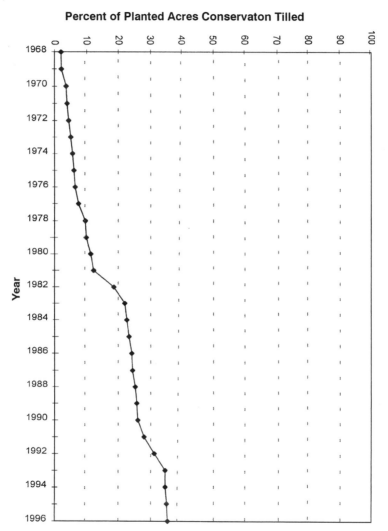

Source: Schertz (1988) and CTIC (annual surveys).

TABLE 2.1a. Trends in Conservation Tillage Use by Crop (Percentage of Planted Acreage Conservation Tilled)

	1989	1990	1991	1992	1993	1994	1995	1996
Corn, dc	48.8	51.6	53.4	50.5	48.1	53.2	53.5	46.4
Corn, f	32.0	32.1	34.7	38.9	43.4	40.4	41.1	40.1
Cotton	3.2	4.8	6.1	7.6	10.0	10.7	10.0	10.3
Fallow	27.1	6.3	28.8	32.0	21.0	24.2	35.4	36.7
Forage	23.1	22.5	20.9	22.0	22.4	24.2	23.4	25.0
Grain, fsd	27.7	27.0	28.5	28.5	29.9	31.0	32.2	32.4
Grain, ssd	19.9	20.8	24.0	25.5	28.3	30.8	29.9	30.4
Other	8.0	9.4	10.2	10.7	12.7	15.4	14.9	15.2
Permanent pasture	42.2	35.6	37.0	34.8	33.4	40.0	34.8	37.9
Sorghum, dc	37.3	42.5	46.4	46.3	41.7	50.6	43.6	37.1
Sorghum, f	29.7	29.9	32.1	29.1	34.7	34.8	36.0	32.2
Soybeans, dc	55.3	27.5	62.8	63.4	64.7	66.3	69.6	69.6
Soybean, f	26.8	27.2	30.5	38.9	47.2	46.3	48.6	49.0
Average	26.0	23.9	28.3	31.4	33.8	34.1	35.5	35.9

Source: Conservation Technology Information Center (annual surveys).

On crop types: dc = double crop
f = full season crop
fsd = fall seeded
ssd = spring seeded

11

TABLE 2.1b. Trends in Conservation Tillage Use by Crop (Million Acres)

	1989	1990	1991	1992	1993	1994	1995	1996
Corn, dc	0.3	0.3	0.3	0.4	0.4	0.4	0.5	0.6
Corn, f	23.1	23.5	25.6	30.1	31.8	31.3	29.3	31.4
Cotton	0.4	0.6	0.8	0.9	1.4	1.5	1.6	1.5
Fallow	6.8	7.5	7.8	2.8	5.1	6.5	9.0	8.4
Forage	1.6	1.6	1.6	1.8	1.8	2.0	1.9	1.8
Grain, fsd	14.2	14.9	14.8	14.6	15.4	15.6	15.7	16.3
Grain, ssd	8.1	8.0	8.8	9.4	10.1	10.9	10.3	11.1
Other	1.7	2.0	2.4	2.4	3.1	3.9	3.8	3.7
Permanent pasture	1.8	2.2	1.5	1.4	1.4	1.4	1.2	1.0
Sorghum, dc	0.3	0.2	0.2	0.2	0.2	0.2	0.2	0.2
Sorghum, f	3.4	3.2	3.3	3.5	3.5	3.2	3.1	3.5
Soybeans, dc	3.4	3.9	4.1	3.7	3.5	3.8	3.9	4.1
Soybean, f	15.2	15.0	17.1	21.5	26.0	26.5	28.6	29.7
Total planted acres	71.7	73.2	79.1	88.7	97.2	99.3	98.9	103.8

Source: Conservation Technology Information Center (annual surveys).

On crop types: dc = double crop
 f = full season crop
 fsd = fall seeded
 ssd = spring seeded

The decline in the use of conservation tillage on sorghum is not easily explainable. Full-season corn is the most extensively grown crop in the United States, accounting for 27 percent (31.4 million acres) of total planted acreage in 1996. Currently, approximately 40 percent of planted acreage on which corn is grown is conservation tilled. Between 1995 and 1996, a reduction in the use of conservation tillage occurred in Indiana, which produces relatively large amounts of corn,[6] and to a lesser degree in Ohio. Both states encountered unusually wet weather conditions at the time of spring planting. Because of that, more than 650,000 acres in Indiana alone (over 5 percent of total planted acreage) that were conservation tilled in 1995 reverted to reduced tillage and/or conventional tillage in 1996 (see Table 2.2). Cotton has the lowest proportion of conservation tillage, increasing from 3 to 10 percent between 1989 and 1996. Producers of other important crops, such as peanuts, potatoes, beets, tobacco, and vegetables, have also adopted improved residue management and erosion control practices, even though their production requirements preclude the use of conservation tillage.

Kentucky leads the nation in conservation tillage, with an adoption rate of 73 percent; Delaware, Iowa, Maryland, Missouri, Nebraska, and Tennessee all have between 50 and 63 percent of cropland conservation tilled (see Table 2.2). States with adoption rates less than 10 percent include Arizona, Florida, Massachusetts, Rhode Island, and Vermont. The Appalachian and Corn Belt regions lead in conservation tillage adoption with 48 and 46 percent, respectively, while the Delta, Southeast, Pacific, and Southern Plains have only between 18 and 23 percent (see Table 2.3).

The substantial decline in the use of conservation tillage in some of the midwestern states between 1993 and 1994 reflects the heavy rains and flooding in 1993 that destroyed crop acreage. Nearly 5 million fewer acres were planted in 1993 than in 1992 and 1994. Most of this land was returned to production in 1994, but rills and gullies on the surface and sand and soil deposits on the bottomlands forced farmers to till the soil more. Much of the decline in mulch tillage in 1994 is attributed to this (see Table 2.4). Thus, the use of conservation tillage in Illinois fell 16.3 percent between 1993 and 1994, while Kansas, Minnesota, Ohio, and Wisconsin experienced similar, although less precipitous, declines.

TABLE 2.2. Trends in Conservation Tillage Use by State (Percent of Planted Acreage Conservation Tilled)

	1989	1990	1991	1992	1993	1994	1995	1996
AL	14.9	12.3	13.4	18.1	19.6	20.1	23.0	26.3
AR	8.5	11.4	9.6	9.6	11.9	13.1	13.2	15.2
AZ	0.8	1.0	1.0	1.0	2.8	4.2	4.5	5.3
CA	7.1	7.0	7.4	14.4	13.2	16.6	16.0	17.1
CO	20.8	17.4	27.3	20.6	22.8	26.1	22.9	31.0
CT	14.1	14.9	11.9	18.9	18.2	16.0	16.9	16.9
DE	53.0	36.2	66.3	58.6	66.0	61.7	62.3	62.6
FL	15.7	5.1	4.1	2.4	5.6	8.4	8.4	9.3
GA	20.3	20.1	17.2	18.2	18.4	18.4	15.4	15.6
IA	29.5	33.0	31.9	43.7	50.4	51.0	52.2	50.5
ID	26.2	24.2	26.5	26.6	30.2	31.1	32.6	30.4
IL	36.7	35.4	42.5	48.0	53.6	37.3	39.8	40.5
IN	42.2	22.2	30.1	36.5	40.6	46.5	50.5	43.5
KS	29.4	22.2	31.7	31.1	36.7	34.0	36.0	35.7
KY	59.3	42.8	59.8	62.8	67.2	67.2	69.1	72.5
LA	7.1	10.6	13.6	14.0	14.6	17.7	16.4	20.1
MA	4.4	6.2	6.1	6.6	6.9	8.3	8.2	8.2
MD	54.5	38.9	51.5	52.5	56.2	58.5	59.5	63.2
ME	40.9	35.4	39.5	17.7	42.6	26.4	22.8	21.4
MI	23.4	28.2	31.2	35.5	41.7	44.8	46.1	48.2
MN	17.6	18.3	19.5	22.1	27.1	23.8	23.5	25.7
MO	35.5	34.8	40.5	43.9	48.0	50.2	50.8	50.1
MS	22.2	15.3	22.2	26.9	28.8	28.6	28.2	28.6
MT	22.4	20.6	28.4	30.0	33.6	38.4	41.5	42.4
NC	18.4	12.4	16.3	17.2	17.4	21.5	23.3	23.5
ND	18.3	17.3	25.2	26.5	28.3	32.7	29.9	29.4
NE	36.9	38.2	44.6	48.6	55.4	57.9	57.7	58.0
NH	9.1	10.4	8.3	7.8	9.0	9.9	13.4	13.4
NJ	26.9	16.2	23.3	30.2	32.8	41.1	35.4	33.8
NM	19.9	12.8	13.5	21.7	28.0	29.4	29.1	30.9
NV	52.2	49.3	49.2	44.6	32.3	32.3	32.3	36.5
NY	17.4	16.3	17.2	19.3	20.6	20.9	20.8	21.2
OH	26.9	29.6	33.7	39.1	49.3	45.7	47.0	45.6
OK	28.1	28.8	28.8	22.4	22.4	22.2	23.0	21.1
OR	33.2	19.9	24.2	22.9	22.9	28.2	24.3	27.7
PA	36.8	35.8	34.9	33.0	36.3	37.0	36.5	38.6
RI	5.4	11.1	8.1	5.3	6.2	3.1	3.2	3.2
SC	16.0	6.8	18.5	19.8	24.0	26.0	26.7	22.3
SD	24.1	22.6	25.7	31.3	33.7	35.6	34.7	38.0
TN	31.1	29.9	37.0	45.0	49.2	57.0	54.6	56.3
TX	20.5	20.7	20.5	20.9	22.8	25.9	25.1	23.8
UT	17.9	13.6	20.2	22.5	23.8	23.4	25.8	30.0
VA	48.7	38.5	50.9	44.4	44.2	45.3	46.4	46.8
VT	3.4	3.6	7.7	4.4	5.3	5.9	5.1	4.4
WA	19.5	12.2	14.5	18.5	18.0	20.8	22.0	22.5
WI	18.1	19.4	21.3	23.8	29.6	28.6	31.3	31.1
WV	33.8	37.0	37.8	39.1	45.1	44.1	44.6	45.9
WY	25.8	13.8	19.9	19.9	17.1	16.4	15.9	26.0

Source: Conservation Technology Information Center (annual surveys).

TABLE 2.3. Trends in Conservation Tillage Use by U.S. Department of Agriculture Regions (Percent of Planted Acreage Conservation Tilled)

	1989	1990	1991	1992	1993	1994	1995	1996
Appalachian	37.4	29.1	38.1	40.8	43.5	45.8	46.7	48.4
Corn Belt	34.1	31.9	36.2	43.4	49.3	45.4	47.4	45.9
Delta States	12.1	12.3	14.2	15.6	17.3	18.7	18.2	20.2
Lake States	18.9	20.5	22.3	25.2	30.6	28.9	29.8	31.4
Mountain	21.7	15.8	25.6	24.4	27.9	29.8	31.3	34.6
Northeast	34.5	29.1	33.8	33.6	36.9	37.8	37.4	38.8
Northern Plains	26.9	24.4	31.7	33.9	37.9	39.2	38.8	39.2
Pacific	16.0	11.5	13.4	17.2	16.3	20.2	19.6	21.1
Southeast	17.5	13.1	13.6	15.1	17.2	18.9	18.3	18.4
Southern Plains	22.8	23.2	22.9	21.3	22.7	24.9	24.5	23.1

Source: Conservation Technology Information Center (annual surveys).

Regional Composition:

Appalachin (WV, KY, NC, TN, VA)
Corn Belt (IA, MO, IL, IN, OH)
Delta States (AR, LA, MS)
Lake States (MN, WI, MI)
Mountain (ID, MT, WY, NV, UT, CO, AZ, NM)
Northeast (ME, PA. NH, CT, NJ, NY, MD, MA, RI, DE, VT)
Northern Plains (ND, SD, NE, KS)
Pacific (WA, OR CA)
Southeast (AL, GA, SC, FL)
Southern Plains (OK, TX)

15

TABLE 2.4. Trends in Conservation Tillage Use by Type

	Type As a Percent of Total Crop			Type As a Percent of Total Conservation Tillage		
	no till	ridge till	mulch till	no till	ridge till	mulch till
1989	5.5	0.9	17.4	21.3	3.4	67.0
1990	6.5	1.0	16.9	27.1	4.1	70.6
1991	7.5	1.0	17.4	26.5	3.7	61.4
1992	9.9	1.1	19.0	31.7	3.6	60.5
1993	12.0	1.1	20.7	35.4	3.3	61.3
1994	13.2	1.1	19.8	38.8	3.3	57.9
1995	14.4	1.1	20.0	40.6	3.1	56.3
1996	14.5	1.1	20.2	40.5	3.0	56.5

Source: Conservation Technology Information Center (annual surveys).

Another important factor leading to the decline in conservation tillage in the Corn Belt in 1994 was the absence of a government set-aside program. Previously idle acres that were returned to production were tilled using conventional practices (U.S. Environmental Protection Agency, 1996).

The largest regional growth in the use of conservation tillage between 1989 and 1996 occurred in the Northern Plains, the Corn Belt, and the Lake States (see Table 2.3). This growth was mostly a function of conservation compliance (Williams et al., 1989; Esseks and Kraft, 1993; Hyberg and Johnston, 1997). In the Northern Plains, recent increases in the use of conservation tillage reflect the rise in the use of such systems to plant and manage small grains (e.g., wheat) as well as corn.

The largest increase in the use of conservation tillage in the past few years has occurred in South Dakota. In 1996 alone, more than 1 million additional acres were conservation tilled. These additional acres were not exclusively switched from conventional tillage to conservation tillage. Planted acreage in South Dakota increased by approximately 1.5 million acres between 1995 and 1996, while the number of fallow acres fell by more than 500,000. It appears that the main reason fallow acres decreased in South Dakota was because of the additional moisture that was retained as a result of using no tillage. This allowed farmers to plant a crop (principally soybeans) instead of leaving the land fallow (Beck, 1996).

Mulch tillage (see Figure 2.1) continues to be the dominant type of conservation tillage (see Table 2.4), although the use of no tillage has increased. Mulch tillage accounted for 57 percent of conservation tillage and was used on 20 percent of the nation's cropland in 1996. No tillage accounted for 41 percent of conservation tillage, being used on 15 percent of the nation's cropland, while ridge tillage was used on only about 1 percent of the cropland in 1996. Over the past few years, a slight increase has occurred in the use of no tillage relative to mulch tillage.

Results from the *Cropping Practices Survey* can provide insights into the use of conservation tillage unavailable from the *National Crop Residue Management Survey* of the CTIC.[7] The CPS focuses on only a limited number of agricultural commodities, but it does include sufficient information to estimate the use of conservation tillage by land classification, that is, highly erodible land (HEL) versus nonhighly erodible land (NONHEL). Figure 2.3 plots the use of conservation tillage on HEL and NONHEL for corn, soybeans, and all wheat (winter, spring, and durum) from 1989 to 1995.[8,9] The use of conservation tillage exhibits an identifiable upward trend until 1993. The trend in the use of conservation tillage on NONHEL emulates that on HEL. Conservation tillage, however, is used about 8 to 10 percent more extensively on HEL than on NONHEL. When total planted acreage for major field crops is considered, about 28 percent of conservation tillage occurs on HEL, while almost 72 percent occurs on NONHEL. Therefore, conservation compliance is not the sole force motivating the adoption of conservation tillage, even though conservation tillage has been most vigorously promoted for its ability to control soil erosion. Economic factors are clearly impacting farmers' decisions to adopt and use conservation practices (*Triazine Network News,* 1996).

For corn and soybeans, the growth in the use of conservation tillage has been greater on HEL than on NONHEL, while wheat has shown little aggregate change over the period 1989 to 1995 (see Tables 2.5 and 2.6). Approximately 15 percent of corn acres surveyed in the *Cropping Practices Survey* in 1995 was designated as HEL, with 60 percent of that acreage planted using conservation tillage. Nearly 14 percent of the land on which soybeans were grown was designated as HEL, and 68 percent of that land used conservation tillage in 1995. Only about one-fourth of wheat planted acreage used conservation tillage in 1995, although over 38 percent of the cropland was designated as HEL.

The CPS also includes questions about the length of time production practices have been used. The responses indicate that most farmers have been using conservation tillage for a relatively short period of time. In 1995, for example, corn farmers who used conservation tillage had been using it, on average, for only 1.5 years. For soybean and wheat farmers, the averages were 2.3 and 0.9 years, respectively. Thus, farmers do not have extensive experience with the management problems for these crops that arise from using conservation tillage.[10] This, in part, can help explain why Indiana farmers, when confronted with wet soil conditions in 1996, reverted to well-known conventional tillage practices.

FIGURE 2.3. Conservation Tillage Use on Major Field Crops by Land Classification: 1989-1995

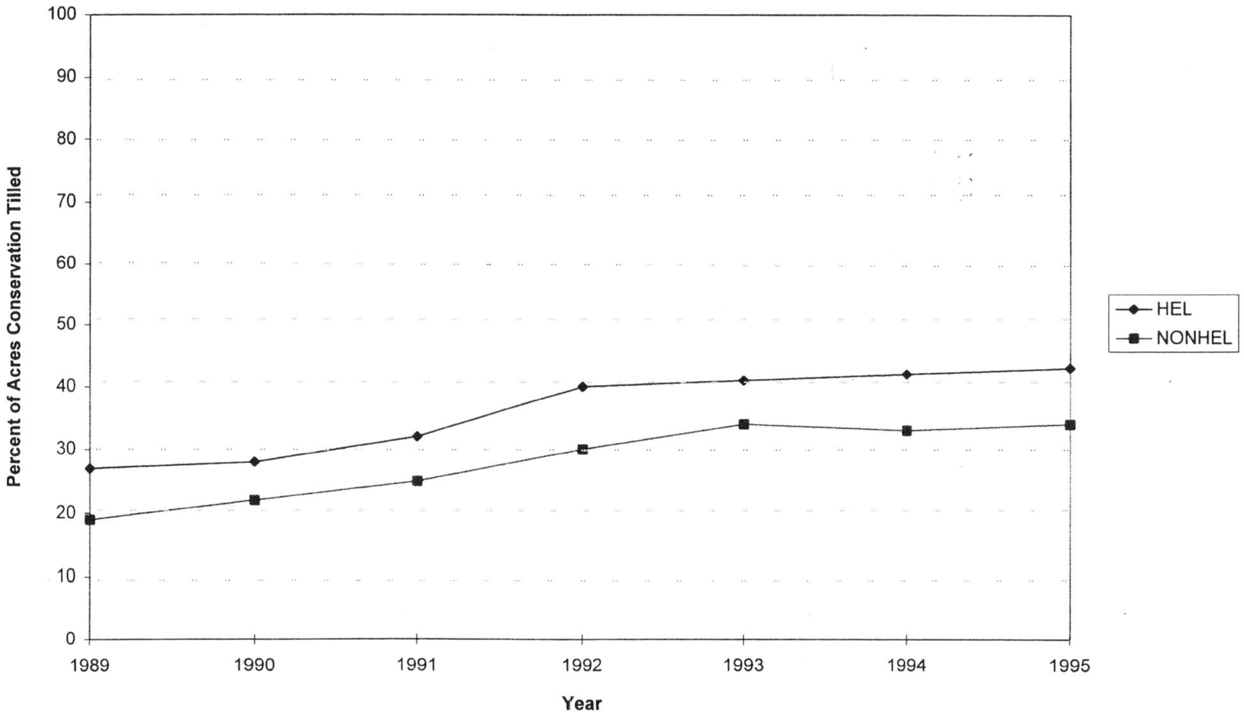

Source: U.S. Department of Agriculture (USDA), Economic Research Service (annual surveys).

TABLE 2.5. Tillage Systems Used in Field Crop Production on Highly Erodible Land, 1989-1995

Crop/Tillage Type	1989	1990	1991	1992	1993	1994	1995
Corn—planted acres (1,000,000)	10.5	12.7	13.3	12.5	11.3	11.9	11.0
Tillage System (percent of planted acres)							
Conventional	67	68	60	49	44	44	40
Conservation Tillage	33	32	40	51	56	56	60
Soybeans—planted acres (1,000,000)	6.3	8.3	8.3	7.9	8.4	8.6	8.8
Tillage System (percent of planted acres)							
Conventional	72	67	55	46	37	35	32
Conservation Tillage	38	33	45	54	63	65	68
Wheat—planted acres (1,000,000)	10.0	14.1	13.0	14.7	16.4	15.7	16.1
Tillage System (percent of planted acres)							
Conventional	76	76	82	73	75	77	74
Conservation Tillage	24	24	18	27	25	23	26

Source: USDA, Economic Research Service (annual surveys).

TABLE 2.6. Tillage Systems Used in Field Crop Production on Non–Highly Erodible Land, 1989-1995

Crop/Tillage Type	1989	1990	1991	1992	1993	1994	1995
Corn – planted acres (1,000,000)	41.0	43.2	44.5	46.9	43.4	48.6	44.8
Tillage System (percent of planted acres)							
Conventional	80	75	72	64	61	61	63
Conservation Tillage	20	25	28	36	39	39	37
Soybeans – planted acres (1,000,000)	39.3	36.6	38.7	37.9	41.5	42.7	43.9
Tillage System (percent of planted acres)							
Conventional	78	75	68	63	55	54	52
Conservation Tillage	22	25	32	37	45	46	48
Wheat – planted acres (1,000,000)	36.5	40.2	35.0	40.0	37.9	37.1	35.9
Tillage System (percent of planted acres)							
Conventional	86	83	86	82	85	87	80
Conservation Tillage	14	17	14	18	15	13	20

Source: USDA, Economic Research Service (annual surveys).

The crop residue left on the field is, not surprisingly, substantially greater for conservation tillage than for conventional tillage. For example, in 1995, the crop residue left after the corn harvest and after tillage on conventional tilled cropland that was conservation tilled was 55.3 percent versus 13.2 percent for conventional tillage. On HEL, the residue cover was somewhat higher, at 65.0 percent for conservation tilled cropland. For soybeans and wheat, the percentages of residue cover were 63.0 and 45.0 percent, respectively, versus 16.5 and 11.3 percent for conventional tillage. On conservation tilled HEL alone, the values were 71.0 and 47.1 percent residue cover for soybeans and wheat, respectively. The relative amounts of crop residue left on the field for these crops has remained roughly constant over the period 1989 to 1995.

FACTORS AFFECTING CONSERVATION TILLAGE ADOPTION AND USE

Farmers, in general, tend to make production practice changes slowly. The adoption process generally can be viewed as having five stages (Nowak and O'Keefe, 1992). Initially, farmers are unaware of a new practice (Stage 1). They become aware of new practices through various sources, including neighbors, farming publications, mass media, County Extension Service agents, chemical dealers, and crop consultants (Stage 2). Farmers then evaluate the practice in terms of their own operation through educational sources such as demonstration projects, talking with agents, and talking with neighbors who have tried the practice (Stage 3). Farmers may then test the practice on part of their farm (Stage 4). The ability of a practice to be tested on part of the farm enhances its potential for adoption (Office of Technology Assessment, 1990; Nowak and O'Keefe, 1995). Finally, full adoption occurs if the practice is found to be better than what they are currently doing (Stage 5).

A variety of economic, demographic, geographic, and policy variables have been identified that affect the adoption and use of conservation tillage in the United States.[11] The rate of adoption (diffusion) of a new technology—for example, conservation tillage—determines the rate of technological change. The first empirical assessment of the diffusion of a new technology was applied to hybrid corn (Griliches, 1957). The diffusion follows an innovation cycle. The cycle starts with efficient producers first introducing the new technology that requires a threshold level of technical skill for profitable use. As skill levels of other farmers increase through experience, the new technology is more widely adopted. The time path of adoption of the new technology can be analytically derived as a function of the distribution of technical ability among producers and the rate of change in technical skill (Kislev and Schori-Barach, 1973). Adoption is also a function of exogenous factors, and these factors will retard or accelerate the rate of adoption. Investment costs associated with the adoption of the new technology will have an important influence on a farmer's choice. Government policy in the form of conservation compliance is an example of an exogenous factor that would be expected to accelerate the rate of adoption of conservation tillage (Batte, 1993; Batte, Forster, and Bacon, 1993). Yet another exogenous consideration is what is nominally referred to as "learning by doing" (Alchien, 1959; Rosen, 1972; Dudley, 1972). When specialized management skills are required for production, owners/operators will gain proficiency with experience; that is, they learn by doing.

Central to the question of adoption of a new technology is the issue of heterogeneity: Why do farmers differ in their willingness to adopt a new technology? Much of the literature concerning the adoption of new technologies focuses on this issue (Antle and McGuckin, 1993; Westra and Olson, 1997). Among the reasons suggested for differences among farmers in the willingness to adopt conservation tillage are entrepreneurial ability, risk preferences, and the availability of complementary inputs (Feder, Just, and Zilberman, 1985). The entrepreneurial or managerial requirements of conservation tillage are quite complex relative to conventional tillage. Management skills, in fact, dominate the successful use of conservation tillage (U.S. Department of Agriculture [USDA], 1997). The adoption of soil-conserving tillage systems normally requires a higher level of management skills to carry out the proper timing and placement of nutrients and pesticides to be successful. Conservation tillage allows for less opportunity to correct mistakes or adjust to changed circumstances once the growing season is underway. The adoption of a new technology by a farmer always involves a degree of risk and uncertainty concerning its impact on output. The risk associated with the adoption of a new technology has been shown to be multifaceted. Just and Zilberman (1983) argued that producers with large farms are more likely to adopt new technologies because of diversification and that the willingness to adopt new technologies depends on the similarity in the inputs between the existing and the new technologies. Finally, new technologies must be integrated into the availability of existing inputs. In the case of conservation tillage, this implies that any conservation tillage system must be compatible with, for example, the soil characteristics and climatic conditions (Nowak, 1984, 1992).

In the context of the diffusion of conservation tillage as a new technology (production practice), a sizable number of studies are available that provide some insights into the important factors that affect its adoption. Because there is considerable redundancy in the results of the studies, an exhaustive survey will not be provided. Pagoulatos and colleagues (1989), using an erosion damage function analysis for corn grown in Kentucky, found that the decision to convert to conservation tillage from conventional tillage is dependent on the price of output, the discount rate (with a higher discount rate leading to a slower adoption of conservation tillage), and the capital cost of conversion. Large capital costs for new machinery serve as a deterrent to adoption of conservation tillage.

Uri (1997) used a two-stage decision model econometrically estimated for corn produced in the United States in 1987. The data came from the Farm Costs and Returns Survey (FCRS) conducted by the U.S. Department of Agriculture.[12] He found that cash grain enterprises were more likely to adopt conservation tillage than other farm types. The slope of the cropland was an important factor (the greater the slope,[13] the greater the likelihood of conservation tillage adoption), and average rainfall (but not average temperature) was associated with a greater likelihood that conservation tillage was adopted. A number of factors, including expenditures on some inputs and farm owner/operator characteristics, were found not to be associated with the adoption or nonadoption of the conservation tillage production practice. For example, the age and education level of the farmer/operator was not associated with the adoption of conservation tillage. The productivity of the soil, as measured by average yield across farms in a county, had no identifiable impact on the decision to adopt conservation tillage. The texture of the soil, the total acres planted, the number of acres in the acreage reduction program, the extent of irrigation, and the proportion of acres not receiving any pesticide treatment likewise were not associated with the adoption of conservation tillage on corn acreage.

Gray and colleagues (1996) used a simulation model to compare the adoption of conservation tillage systems to conventional tillage systems for wheat production in western Canada. Crop yield and the price of the burndown herbicide (the herbicide used to eliminate vegetation prior to planting) were key determinants to the short run profitability of adopting conservation tillage. The price of fuel was also important, although less so.

Carter and Kunelius (1990), analyzing data from Atlantic Canada, found that some soil types are simply not suitable for conservation tillage.[14] Certain soils require a high degree of cultivation to maintain their structure and regular tillage to ensure adequate crop productivity. Moreover, climatic constraints such as a short growing season, cold temperatures, and excessive precipitation can influence the choice of a conservation tillage system.

The greater risk associated with the use of conservation tillage has been shown to be a deterrent to its adoption in a number of studies. Risk in these studies is typically defined as variability in yields or variability in net returns. Thus, Mikesell and colleagues (1988), using a simulation model, evaluated the expected net returns and risk of alternative tillage systems for a 640-acre grain farm in northeastern Kansas. Conservation tillage systems had slightly higher expected incomes but were more risky. A risk-averse farmer would prefer conventional tillage to conservation tillage. Williams and colleagues (1988), using a simulation model, found that conservation tillage used in grain sorghum production had higher expected net revenues but greater risk than conventional tillage.

Westra and Olson (1997) estimated a structural model based on survey responses for farmers in two counties in Minnesota. Their results suggest that larger farms are more likely to use conservation tillage. Also, if the owner/operator is relatively more concerned about erosion, the probability of adopting conservation tillage is greater. The greater complexity associated with the use of conservation tillage requiring greater management skill is identified as a deterrent to the adoption of conservation tillage.

The consensus of the studies cited here, plus others,[15] is that the relative economic performance of any conservation tillage practice depends on a number of site-specific factors. The degree to which farmers are risk averse, soil type, topsoil depth, choice among cropping systems, level of managerial expertise, and local climatic conditions have all been identified as important variables. Consequently, a farmer's decision to use conservation tillage will depend on these site-specific and operator-specific factors. It is the aggregate effect of these site-specific factors superimposed on operator characteristics and the basic diffusion model for a new technology that has contributed to the evolution of the trend in conservation tillage use in the United States. Site-specific factors impact both the rate of adoption of conservation tillage and the ultimate extent of its use.

CONCLUSION

The adoption of conservation tillage in the United States experienced a pronounced upward trend until 1993. Since that time, the percentage of the planted acreage using conservation tillage practices has exhibited a very slight upward trend. One of the challenges in examining the trend in conservation tillage use is to explain its evolution in terms of economic and geographic factors and government policy. Although it is difficult to precisely quantify the impact of any specific factor, a number of empirical studies suggest that different economic and geographic variables do affect decisions to adopt conservation tillage and, ultimately, the extent of its use. Relative performance—net private benefits—is what matters. Consequently, any attempt to increase the adoption of conservation tillage should address this issue.

The following chapter examines the economic and environmental effects of more extensive use of conservation tillage in the United States.

Chapter 3

Benefits and Costs
of Conservation Tillage

- The use of every production system, including conservation tillage, has economic and environmental consequences. A system designed to satisfy environmental goals cannot be fairly evaluated on private economic criteria alone.
- The economic benefits of the adoption of conservation practices depend on site-specific factors, including soil characteristics, local climatic conditions, cropping patterns, and other attributes of the overall farming operation. Although it is possible to draw some general inferences about components of economic returns and costs, a comprehensive assessment of the net private benefits from greater use of conservation tillage is not feasible.
- Switching the remaining 22.4 million acres of highly erodible land that is currently conventionally tilled to conservation tillage will increase social benefits by about $49.6 million annually.
- Nonquantifiable social benefits associated with conservation tillage adoption include improved wildlife habitat and reduced atmospheric emissions.

In response to growing concerns about the impact of agricultural production on the environment, the use of conservation tillage has increased. Use of these practices frequently reduces surface water contamination and may enhance soil quality. Because many of the conservation tillage practices have been designed to attain environmental objectives, these systems cannot be fairly evaluated on productivity criteria (net private benefits) alone. In this chapter, these economic and environmental issues associated with the adoption of conservation tillage will be explored.

Historically, productivity criteria were the sole factors considered in comparing and contrasting alternative production technologies. Since the late 1940s, research on crop production focused on reducing labor requirements and increasing yields per unit of land (Hayami and Ruttan, 1985). Economic evaluations of a new production technology were primarily concerned with the effects of the new technology on profitability. Yield increases by themselves may not raise profits, so the central issue was whether the value of the yield increase justified the costs incurred to obtain it. In evaluating whether to adopt a new practice, a farmer would consider whether net farm income (net private benefits) would increase. Explicit fixed and variable costs would be considered as well as expected prices and yields. A farmer would also include the implicit or subjective costs associated with switching production technologies. Included in the measure of net farm income would be any government program payments contingent upon the adoption of a new production technology such as conservation tillage.[1]

MEASURING THE BENEFITS AND COSTS OF CONSERVATION TILLAGE

Profitability and environmental impact are the two performance criteria of greatest interest for evaluating the impact of conservation tillage adoption. Profitability is the main criterion for private economic (net private benefits) decisions, while the environmental impact criterion compares the net social benefits of the different production practices.

Every production practice, including conservation tillage, has positive or negative environmental consequences that may involve air, land, water, and/or the health and ecological status of wildlife. The negative impacts associated with agricultural production, and the use of conventional tillage systems in particular, include soil erosion, energy use, leaching and runoff of agricultural chemicals, and carbon emissions. As with profitability, it is not solely average environmental effects that are important but also the stability of these effects.

The impacts of conservation tillage will be measured as the change from the conditions associated with the use of conventional tillage practices. Conventional tillage practices typically involve the use of intensive tillage and generally leave less than 30 percent crop residue cover on the soil after harvest. Different types of plows include the moldboard plow, chisel plow, subsoiler, disk plow, off-set disk, and blade plow (Dickey, 1992).[2] Note, however, that because conventional farming practices are continually evolving and vary geographically, the point of comparison must be typical of common practices for the time and location of the assessment. The relative gain from adopting conservation tillage will depend on the system from which the farmer switched.

ECONOMIC BENEFITS AND COSTS TO FARMERS OF CONSERVATION TILLAGE ADOPTION

A farmer who chooses conservation tillage over conventional tillage does so in hopes that it will maximize net farm income (profit) and/or mitigate risk.[3] Consequently, in assessing the economic benefits and costs of conservation tillage versus conventional tillage, it is necessary to evaluate each of the components that contributes to net farm income and risk. These components include yield, expected output prices, and inputs used in the production of agricultural commodities.[4]

Yield

Site-specific factors such as soil characteristics, local climatic conditions, cropping patterns, and other attributes of the overall farming operation influence yield. Conservation tillage can affect soil characteristics such as structure, organic matter content, and soil microbial populations that influence the movement of water in and through the soil, thereby potentially resulting in an increase in yield (Bruce et al., 1995; Ismail, Blevins, and Frye, 1994; Office of Technology Assessment, 1990; Paudel and Lohr, 1996). Additionally, enhanced water infiltration associated with greater surface residue provides additional soil moisture that can benefit crops during low rainfall periods (J. Baker, 1987; Wauchope, 1987).

Yield benefits associated with the continuous use of conservation tillage take a relatively long time to materialize, perhaps a minimum of ten years on some soils and under certain climatic conditions, to realize fully any potential yield effects associated with improved soil characteristics such as soil tilth (Hudson and Bradley, 1995; Ismail, Blevins, and Frye, 1994; Lal, Logan, and Fausey, 1990; Olsen and Senjem, 1996; Zobeck et al., 1995).[5] Moreover, a single soil disturbing tillage activity may eliminate any improvement in soil characteristics that had been realized (University of Illinois Agricultural Extension Service, 1997; Moldenhauer and Black, 1994).[6]

The effect of conservation tillage on yield is not unequivocal. Before elaborating on this, two points need to be made. First, since conservation tillage consists of different subclasses, yield effects will vary not only by site-specific characteristics but also by which conservation tillage practice—mulch tillage, ridge tillage, or no tillage—is used (see Figure 2.1 in Chapter 2). Second, results of field experiments on the effects of conservation tillage on yields are mixed. Several selected studies are reviewed here, and citations are provided for many more.

The seven University of Illinois Agricultural Research and Demonstration Centers have evaluated crop yield response to different tillage systems under a wide variety of soil and climatic conditions (Siemans, 1997). The results have shown that crop yields vary due more to weather conditions during the growing season than to the tillage system used. Corn and soybean yields are generally higher when the crops are rotated than when either crop is grown continuously. Comparative yields due to tillage system vary with soil type (see Table 3.1). Corn and soybean yields have generally been found to decrease slightly as tillage is reduced on poorly drained soils. On these soils, however, the ridge tillage system has often produced higher corn yields. On well to moderately well-drained, medium-textured soils, yields with all tillage systems are similar for rotation corn and soybeans. With continuous corn, yields generally decrease as tillage is reduced. On excessively drained sandy soils, conservation tillage systems typically produce high yields.

Epplin and colleagues (1994) conducted field experiments on wheat over a ten-year period (1977 through 1986), comparing six different tillage systems: moldboard plow, chisel, disk, sweep, and no tillage. Ridge tillage and mulch tillage were not considered.

TABLE 3.1. Corn and Soybean Yields in Illinois by Tillage System and Soil Type

	Soil type, years of experiment						
Tillage system	Thorp silt loam, 1991-1995	Alford silt loam, 1989-1993	Flanagan silt loam, 1990-1994	Cisne silt loam, 1990-1994	Downs-Fayette silt loam, 1990-1994	Tama silt loam,[1] 1990-1994	Tama silt loam,[2] 1988-1994
	Five-year average corn yields (bu/acre)						
Moldboard plow	*	155	163	*	172	165	140
Chisel plow	164	156	153	138	170	155	137
Disk	170	160	160	138	171	165	135
No tillage	164	153	150	133	167	159	136
	Five-year average soybean yields (bu/acre)						
Moldboard plow	42	41	54	28	44	54	*
Chisel plow	*	44	50	29	47	55	*
Disk	43	46	52	*	46	53	*
No tillage	40	42	54	32	45	53	*

Source: Siemans (1997, p. 129).

*Denotes system not included in experiment.

[1] Corn-soybean rotation
[2] Continucus corn

31

Wheat yields from the moldboard plow tillage systems were consistently greater and showed less variability than yields from the three intermediate tillage systems and no tillage. The lowest yield was from the no tillage practice. Moreover, yield was inversely related to the amount of crop residue cover on the field prior to planting. This is attributed to the rootborne and soilborne pathogens, secondary toxins, and increased weed competition associated with the higher crop residue cover.

Young and colleagues (1994) found that, based on six years' worth of field experiment results on wheat grown in eastern Washington, yields increased under conservation tillage relative to conventional moldboard plow tillage. Clark and colleagues (1994) reported that the tillage practice for wheat in conjunction with whether the crop was planted continuously or rotated on a wheat-fallow basis were the critical factors in determining whether conservation tillage resulted in any change in yield relative to conventional tillage practices. In general, when conservation tillage was used in combination with a wheat-fallow rotation, yields increased by about 10 percent.

As noted, many other studies compared the yield effects of conservation tillage to the yields associated with conventional tillage practices. Most of the studies were related to specific crops or cropping patterns and were limited in their geographic extent (see the various volumes in the *Crop Residue Management to Reduce Soil Erosion and Improve Soil Quality* series compiled by the U.S. Department of Agriculture, Agricultural Research Service; Fox et al., 1991; Roberts and Swinton, 1996.) The conclusion from these studies is that no clear consensus emerges concerning the effect of conservation tillage on yield relative to conventional tillage. The factors previously noted, including soil characteristics, climatic conditions, pest pressures, and cropping patterns, result in site-specific variation in yields, making it difficult to extract broad inferences.

One area showing is empirical consistency concerns the effects of conservation tillage on yield variability. It is generally found that yield under conservation tillage was riskier (more variable) than yield under conventional tillage. For example, Williams and Mikesell (1987), using the coefficient of variation as a measure of risk, found that for grain sorghum and soybean production in northeast

Kansas, net returns under conservation tillage were considerably riskier than under conventional tillage. Setia (1987), using an expected utility maximization framework, found that yields for both continuous corn and corn and soybeans in rotation are riskier under conservation tillage than conventional tillage.

Economic Research Service (ERS) *Cropping Practices Survey* data (described in Chapter 2) can be used to investigate the yield/ tillage relationship across different geographic regions and soil types. Estimates of average yields (classified by tillage practice for nonirrigated corn, soybeans, winter wheat, spring wheat, and durum wheat for 1990 to 1995 and shown in Table 3.2) found no statistically significant difference in yields between conventional tillage and conservation tillage.[7] This result carries over when conservation tillage is disaggregated to mulch tillage, ridge tillage, and no tillage, although the sample sizes for these subcategories are relatively small. Similar patterns were observed for irrigated crop production.

Another important result from the CPS data is the lack of a statistically significant difference between the variance in the yield for conservation tillage and for conventional tillage for corn and soybeans, although a difference was observed for the winter and spring wheat. For durum wheat, no statistically significant difference existed between the variance in the yield for conservation tillage and for conventional tillage for 1990 to 1992, but the variance was different for 1993 to 1995. Therefore, risk may be an important factor in the adoption of conservation tillage for wheat farmers, but apparently not for corn and soybean producers.

ERS CPS data can also be used to conduct an analysis of variance to determine which factors have a statistically identifiable impact on yield.[8] Factors considered, in addition to the tillage practice used, include whether the cropland was designated highly erodible land, the number of hours devoted to tillage operations, the residue cover, the number of consecutive years that no tillage has been used (on cropland using conservation tillage), and the previous crop planted. The single factor that is found to significantly affect yield is whether the cropland is designated as highly erodible (see Table 3.3). Yields were lower on HEL. Moreover, this significance was not dependent on the interaction between the HEL designation and any other vari-

TABLE 3.2. Average Yields by Tillage Practice—1990-1995

Commodity/Year	Conventional Tillage	Conservation Tillage
Corn (bu/acre)		
1990	122.1 (1.20)[1]	122.6 (1.39)
1991	114.5 (1.17)	117.1 (1.91)
1992	143.1 (1.33)	146.7 (1.58)
1993	106.5 (0.80)	104.7 (0.87)
1994	146.1 (0.71)	144.9 (0.86)
1995	118.0 (0.95)	116.1 (1.11)
Soybeans (bu/acre)		
1990	38.1 (0.25)	37.7 (0.41)
1991	40.2 (0.60)	39.6 (0.42)
1992	41.7 (0.34)	43.1 (0.39)
1993	37.6 (0.28)	39.2 (0.29)
1994	45.0 (0.26)	45.4 (0.28)
1995	39.9 (0.40)	39.7 (0.36)
Winter Wheat (bu/acre)		
1990	41.1 (0.51)	43.4 (1.04)
1991	41.5 (0.58)	39.1 (0.97)
1992	37.5 (0.49)	38.4 (1.05)
1993	42.4 (0.64)	42.7 (1.04)
1994	41.2 (0.54)	42.5 (1.12)
1995	44.4 (0.76)	44.4 (1.18)
Spring Wheat (bu/acre)		
1990	42.1 (1.06)	39.9 (1.87)
1991	35.7 (0.70)	31.8 (1.98)
1992	46.7 (1.14)	38.7 (2.35)
1993	37.6 (1.21)	32.6 (2.23)
1994	34.1 (0.75)	31.9 (1.33)
1995	33.3 (0.84)	32.5 (1.35)
Durum Wheat (bu/acre)		
1990	40.0 (1.39)	38.7 (1.42)
1991	35.1 (0.91)	31.6 (0.94)
1992	40.7 (1.27)	36.8 (1.35)
1993	35.0 (2.35)	36.0 (5.19)
1994	33.6 (0.84)	33.7 (1.28)
1995	31.2 (0.92)	29.4 (1.35)

Source: USDA, Economic Research Service, *Cropping Practices Survey* (1990-1995, pp. 1-34)

[1]The values in parentheses are the standard errors.

TABLE 3.3. Analysis of Variance of Yields Based on the *Cropping Practices Survey*—1995

Commodity	Variable		
	HEL[1]	CONTIL[2]	HEL *CONTIL[3]
Corn	53,545.6	683.5	3,838.3
	(**)	(ns)	(ns)
Soybeans	7,549.5	641.9	1,529.9
	(**)	(ns)	(ns)
Winter Wheat	12,497.5	278.1	0.0
	(**)	(ns)	(ns)
Spring Wheat	8,601.4	213.1	2,716.6
	(*)	(ns)	(ns)
Durum Wheat	233.5	418.2	89.9
	(ns)	(ns)	(ns)

Source: USDA, Economic Research Service, *Cropping Practices Survey* (1995, p. 12).

Note: The values in the table are the ANOVA sum of squares for the respective variables. The characters in parentheses indicate whether the sum of squares is not statistically significant (ns), whether it is statistically significant at the 5 percent level or better (*), or whether it is statistically significant at the 1 percent level or better (**).

[1] HEL indicates whether the land was designated as highly erodible land.
[2] CONTIL indicates whether conservation tillage was used.
[3] HEL *CONTIL is the interaction term.

able. Thus, the difference in yield on HEL versus NONHEL land was not contingent on, for example, the tillage practice used or the previous crop grown or the use of conservation tillage in conjunction with the fact that the cropland is designated HEL. This holds for corn, soybeans, and the different types of wheat. Finally, the resulting statistical significance of just HEL does carry over to the years 1990 to 1994.

In sum, for the crops for which survey data are available—corn, soybeans, and wheat (winter, spring, and durum)—no statistically identifiable association exists between tillage practice and yield for the years 1990 to 1995; yields are neither higher nor lower. There is, however, greater variability in the yields for winter wheat, spring wheat, and durum wheat that are conservation tilled.

Production Costs

For evaluating the profitability of conservation tillage relative to conventional tillage, the related costs of production are an important consideration. The different tillage practices may affect the cost of labor, fertilizers, pesticides, seed, and machinery. Grain handling and drying costs are affected if yields differ. Land cost is normally assumed not to vary with the tillage system.

Labor Use and Cost

A reduction in the intensity and number of tillage operations lowers costs for labor and machinery, especially if the machinery is used optimally (Siemans and Doster, 1992). Several studies estimate the savings in labor costs if conservation tillage is adopted. Weersink and colleagues (1992), for example, found that corn-soybean farmers in southern Ontario realized significant savings in labor costs with no tillage and ridge tillage compared with conventional tillage systems. The omission of preplant operations alone reduces labor requirements by up to 60 percent.

Dickey and colleagues (1992) calculated the typical labor requirements from machinery management data for various tillage systems in Nebraska (see Table 3.4). The moldboard plow system has the greatest labor requirement for tilling and planting corn and soybeans. Compared with the commonly used disk system, no tillage, for example, saves about twenty minutes of labor per acre.

ERS *Cropping Practices Survey* data also show that labor savings can be significant. The number of hours devoted to tillage operations are different between conservation tillage and conventional tillage, but the relative amount of time spent on tillage operations has not changed appreciably over time. In 1995, conventional tillage operations for corn took 0.38 hour per acre, while conservation tillage required only 0.19 hour per acre. Soybeans and wheat took, respectively, 0.44 and 0.47 hour per acre for conventional tillage, while conservation tillage operations on average took 0.20 and 0.22 hour per acre. All of these differences are statistically signifi-

TABLE 3.4. Labor Requirements for Various Tillage Systems in Nebraska

Operation	Moldboard Plow	Chisel Plow	Disk	Ridge Tillage	No Tillage
			Hours per acre		
Chop Stalks	0.38	0.21	na[1]	0.17	na
Fertilize, knife	0.13	0.13	0.13	0.13	0.13
Disk	0.16	0.16	0.16	na	na
Plant	0.21	0.21	0.21	0.25	0.25
Cultivate	0.18	0.18	0.18	0.36	na
Spray	na	na	na	na	0.11
Total	1.22	0.89	0.84	0.91	0.49

Source: Dickey, Jasa, and Shelton (1992, p. 89).

[1]na denotes not applicable.

37

icant. Moreover, the relative amount of time devoted to the tillage operations for conservation tillage and conventional tillage has not changed between 1990 and 1995.

The benefit from conservation tillage of reduced labor needs is greater than just the labor cost savings per acre. There is also the associated opportunity cost of the labor and time saved. That is, if less hired labor is needed, there will be direct savings. Saving the farmer's or other family labor may permit them to engage in off-farm activities. Lower labor requirements for tillage lead to additional returns from the expansion of existing enterprises or allow time for new activities to improve profitability for the whole farm operation.

Fertilizer Use and Cost

The determination of fertilizer needs to attain optimum yields depends on an accurate assessment of the soil's available nutrients in relation to the needs of the crop. This assessment will include site-specific factors such as soil characteristics, cropping patterns, and climatic conditions in addition to the tillage practice. Conservation tillage, however, requires improved fertilizer management (Halvorsen, 1994; Rehm, 1995). In some instances, increased application of specific nutrients may be necessary and specialized equipment may be required for proper fertilizer placement, thereby contributing to higher costs (Griffith and Mannering, 1974).

The results of field experiments indicate that an increase in the amount of crop residue cover on the soil surface tends to keep soils cooler, wetter, less aerated, and denser (Mengel, Moncrief, and Schulte, 1992). These characteristics and beneficial impacts from increased organic matter, improved moisture retention and permeability, and reduced nutrient losses from erosion are associated with conservation tillage and can affect the ability of crops to utilize nutrients. With higher levels of crop residue, proper timing and placement of nutrient applications are critical to enhance fertilizer efficiency to achieve optimal yield at lowest cost (Bosch, Fuglie, and Keim, 1994).

The overall effect of conservation tillage on fertilizer use is subject to some disagreement. Halvorsen (1994), for example, suggests that fertilizer requirements, and hence use, are the same under conservation tillage and conventional tillage. Mengel and colleagues (1992) argue that it is not possible to determine a priori what will happen to fertilizer use because use is a function of many site-specific factors. Finally, Rehm (1995) suggests that fertilizer use will actually fall under conservation tillage because better fertilizer management practices (e.g., injection) will be used.

The *Cropping Practices Survey* can also be employed to examine the relationship between tillage and fertilizer use. For the most part, fertilizer use on conservation tilled acreage was not statistically significantly different from fertilizer use on conventional tilled acreage in 1995 (see Table 3.5). The exceptions are for potash used in corn, winter wheat, and spring wheat production. The reason for this is unknown. A similar pattern was observed for the years 1990 to 1994.

The CPS data, coupled with site-specific climatic information and fertilizer prices, allow for the estimation of demand functions for fertilizer. Demand functions are estimated for nitrogen, phosphate, and potash and are used to identify the factors, in addition to tillage practice, that affect fertilizer use. Although there is some variability across crops in the estimation results, use of both nitrogen and phosphate is about the same for conservation-tilled crops as for conventional-tilled crops. That is, fertilizer use, and hence costs, in the aggregate are the same for the different tillage practices although site-specific/crop-specific/non-tillage-related practice-specific variation will exist.

Finally, the number of trips across a field to apply fertilizer did not vary between tillage systems. The average for corn for 1995 was 1.92 (0.03)[9] trips per field for conservation tillage and 1.90 (0.02) trips per field for conventional tillage. The average for soybeans, winter wheat, spring wheat, and durum wheat for conservation tillage is 0.31 (0.02), 1.32 (0.05), 1.31 (0.09), and 1.46 (0.11), respectively. The average for soybeans, winter wheat, spring wheat, and durum wheat for conventional tillage is 0.32 (0.02), 1.40 (0.03), 1.29 (0.05), and 1.40 (0.09), respectively. Consequently, no statistically significant differences in labor costs or machinery operating and maintenance expenses are associated with the application of fertilizer between conservation tillage and conventional tillage. The same results also hold for 1990 to 1994.

TABLE 3.5. Average Fertilizer Use by Tillage Practice—1995

Commodity	Conventional Tillage	Conservation Tillage
	Pounds per acre	
Corn		
Nitrogen	130.8 (1.92)[1]	136.5 (3.21)
Phosphate	47.9 (1.22)	43.9 (2.30)
Potash	59.3 (1.74)	49.6 (1.79)
Soybeans		
Nitrogen	4.5 (0.74)	4.5 (0.64)
Phosphate	13.2 (1.36)	10.4 (1.00)
Potash	22.5 (1.95)	22.5 (1.87)
Winter Wheat		
Nitrogen	61.7 (1.61)	55.6 (2.84)
Phosphate	20.9 (0.87)	23.5 (1.84)
Potash	9.3 (0.87)	17.0 (2.08)
Spring Wheat		
Nitrogen	58.0 (2.91)	47.4 (4.93)
Phosphate	26.2 (1.42)	29.9 (1.52)
Potash	6.23 (0.91)	1.37 (0.62)
Durum Wheat		
Nitrogen	59.9 (4.55)	65.1 (5.88)
Phosphate	17.6 (1.55)	19.0 (2.09)
Potash	1.3 (0.49)	1.5 (1.49)

Source: USDA, Economic Research Service, *Cropping Practices Survey* (1990-1995, p. 19).

[1]Values in parentheses are the standard errors.

Pesticide Use and Cost

Weed control problems vary among tillage systems because the nature of the weed population changes. Tillage prepares a seedbed not only for the crop but for weed seeds as well (Monson and Wollenhaupt, 1995). Different weed species occur as tillage is reduced, requiring different control programs. Effective weed control with herbicides depends on spraying at the right stage of plant growth, plant stress, weather conditions, and so on. Weed growth and development, as well as appropriate management strategies, vary with location. A weed management program must be site specific and circumstance specific and will be different between conservation tillage and conventional tillage (Martin, 1992).

The reservoir of dormant weed seeds resident in the soil will not be transferred to the germination zone near the soil surface by tillage. Consequently, as annual weeds are controlled, the overall weed problem may decrease after a few years when fields are converted to conservation tillage and if effective weed control is practiced (Fawcett, 1987).

The *Cropping Practices Survey* can shed some light on the relative use of pesticides across different tillage systems. The survey results show that herbicide application rates for conservation tillage were slightly greater than for conventional tillage over the period 1990 to 1995 for corn, soybeans, and winter wheat (see Table 3.6). During this period, conservation-tilled acreage more than doubled. The pattern observed is consistent with the suggestion that in the first few years with a no tillage system, farmers often use more herbicides. Other factors such as weather, soil type, tillage system experience, and endemic weed problems are potentially more important factors than the tillage system used in determining herbicide use. Moreover, the relative impacts of these factors vary from year to year.

Less insecticide is used for conservation tillage than for conventional tillage (Bull et al., 1993). From the CPS for 1995 for corn, for example, insecticide use on conservation-tilled acreage was 0.61 (0.03)[10] pounds per acre and 0.78 (0.04) pounds per acre for conventional-tilled acreage. For soybeans, insecticide use was 0.37 (0.05) and 0.50 (0.08) pounds per acre for conservation-tilled and conventional-tilled acreage, respectively. The lower amounts of insecticide use on soybeans than on corn reflects a greater use of crop rotations.

Results from the ERS *Cropping Practices Survey* suggest that the number of pesticide treatments is greater for corn, soybeans, and spring wheat grown under conservation tillage than under conventional tillage for some years between 1990 and 1995 (see Table 3.7). One possible reason for this is that pest problems occasionally are greater under conservation tillage. This is consistent with a recent

TABLE 3.6. Average Herbicide Use by Tillage Practice

Commodity	Conventional Tillage	Conservation Tillage
Corn	Pounds of active ingredient applied per acre	
1990	3.38 (1.01)[1]	3.27 (1.06)
1991	2.98 (0.71)	3.25 (0.59)
1992	3.03 (0.66)	3.32 (0.68)
1993	2.99 (0.53)	3.42 (0.48)
1994	2.77 (0.51)	3.33 (0.44)
1995	2.72 (0.89)	3.31 (0.63)
Soybeans		
1990	1.28 (0.96)	2.09 (1.21)
1991	1.22 (0.68)	1.51 (0.51)
1992	1.14 (0.50)	1.31 (0.49)
1993	1.06 (0.65)	1.38 (0.48)
1994	1.10 (0.51)	1.35 (0.30)
1995	1.02 (1.20)	1.35 (0.68)
Winter Wheat		
1990	0.26 (0.47)	0.54 (0.51)
1991	0.29 (0.44)	0.71 (0.41)
1992	0.30 (0.47)	0.32 (0.44)
1993	0.28 (0.30)	0.47 (0.50)
1994	0.31 (0.43)	0.43 (0.50)
1995	0.25 (0.37)	0.36 (0.50)

Source: USDA, Economic Research Service, *Cropping Practices Survey* (1990-1995, p. 26).

[1] Values in parentheses are the standard errors.

survey of farmers in the Midwest, who rate the overall pest control problems to be 11 to 14 percent more severe on conservation-tilled soils than on conventional-tilled soils (Pike, Kirby, and Kamble, 1997). Another possible reason is that perhaps split applications are more prevalent with conservation tillage. The data are insufficient, however, to test this hypothesis.

An important caveat needs to be considered about the relationship between pesticide use and tillage practices. Pesticide use is measured by the number of pounds of active ingredients applied per acre. The active ingredient is the component of pesticide products that kills, repels, attracts, or controls the target pest. Aggregated

TABLE 3.7. Average Number of Pesticide Treatments by Tillage Practice—1995

Commodity	Conventional Tillage	Conservation Tillage
Corn		
1990	1.63 (0.02)[1]	1.69 (0.03)
1991	1.60 (0.02)	1.78 (0.04)
1992	1.71 (0.02)	1.82 (0.03)
1993	1.67 (0.02)	1.77 (0.02)
1994	1.72 (0.02)	1.82 (0.02)
1995	1.79 (0.03)	1.93 (0.03)
Soybeans		
1990	1.49 (0.02)	1.51 (0.03)
1991	1.54 (0.03)	1.63 (0.04)
1992	1.57 (0.03)	1.63 (0.03)
1993	1.47 (0.02)	1.62 (0.02)
1994	1.67 (0.02)	1.77 (0.02)
1995	1.64 (0.03)	1.87 (0.04)
Winter Wheat		
1990	0.41 (0.02)	0.55 (0.05)
1991	0.39 (0.02)	0.36 (0.05)
1992	0.39 (0.02)	0.32 (0.04)
1993	0.50 (0.02)	0.49 (0.05)
1994	0.53 (0.02)	0.55 (0.05)
1995	0.76 (0.02)	0.75 (0.05)
Spring Wheat		
1990	1.26 (0.05)	1.42 (0.14)
1991	1.17 (0.04)	1.20 (0.07)
1992	1.14 (0.05)	1.03 (0.07)
1993	1.19 (0.06)	1.12 (0.08)
1994	1.24 (0.07)	1.18 (0.05)
1995	1.0 (0.02)	1.17 (0.03)
Durum Wheat		
1990	1.48 (0.09)	1.52 (0.12)
1991	1.38 (0.09)	1.47 (0.10)
1992	1.45 (0.11)	1.27 (0.09)
1993	1.67 (0.27)	1.56 (0.33)
1994	1.38 (0.07)	1.42 (0.10)
1995	1.41 (0.09)	1.50 (0.14)

Source: USDA, Economic Research Service, *Cropping Practices Survey* (1990-1995, p. 39).

[1] Values in parentheses are the standard errors.

active ingredient statistics, however, such as for the agricultural sector as a whole, for a crop, for a state, for a tillage system, or for a class of pesticides, often obscure significant differences among different pesticide products. For example, certain products are applied in pounds of active ingredient per acre, while competing products are only applied at ounces per acre for the same crop and same target pest. Although near term year-to-year comparisons can be useful in such instances, aggregate measures of pounds of active ingredient applied over time can be very misleading because of product changes, planted acreage shifts, regional shifts in acreage devoted to different crops, regulatory changes, and pest infestation cycles. For example, recommended application rates for herbicides have declined over time as new products have been introduced. As the analysis moves from a highly aggregated to a more disaggregated level, such as to the individual active ingredient level, poundage comparisons over time, crop, or state are more meaningful. This issue is explored more fully by Barnard and colleagues (1997).

Few definitive statements can be made about pesticide use under conservation tillage versus conventional tillage. More herbicides are typically used during the first few years of conservation tillage. Insecticide use falls with conservation tillage. The existing cropping pattern, however, plays an important role in determining pesticide requirements because monoculture systems generally require greater pesticide use than crop rotation systems (USDA Economic Research Service, 1997).

Seed Use and Cost

Seed cost in the context of conservation tillage has not been extensively studied. The question of whether the seeding rate or the type of seed used varies under conservation tillage relative to conventional tillage needs further study. Most field experiments (e.g., Iowa State University, annual reports) maintain a constant seeding rate when comparing yields from conservation tillage to conventional tillage. A few studies casually considered seed costs. For example, it is has been recommended that when soybeans are drilled or planted in narrow rows, the seeding rate should be increased 10 to 20 percent compared with planting in rows thirty inches or wider (Siemans, 1997). The aggregate effects of these types of recommendations on the actual seeding rate, however, is an empirical issue that is a topic for future research (Aw-Hassan and Stoecker, 1990).

Results of the *Cropping Practices Survey* indicate that farmers did not vary the seeding rate by tillage practice in 1990 to 1995 (see Table 3.8). Additionally, the variability in the seeding rate for each of the years and crops is not statistically significantly different between conservation tillage and conventional tillage.

Empirical results for 1995 found no statistically significant differences between the seeding rate and the tillage practice used, whether the cropland was designated HEL, the number of hours devoted to tillage operations, the residue cover, the number of consecutive years that no tillage has been used (on cropland using conservation tillage), and the previous crop planted (see Table 3.9). The statistical significance carries over to the years 1990 to 1994, as well.

Another question that the ERS CPS can address is the importance of tillage on the use of herbicide-resistant or Bt-enhanced seeds. For corn for 1995 (the only year for which the requisite data are available), there is no indication that farmers using conservation tillage were any more likely to use herbicide-resistant hybrid seed or a Bt-enhanced variety of seed for insect control. Only 6.3 (0.7)[11] percent of conventional tillage farmers used a herbicide-resistant seed in 1995, while just 7.6 percent (0.9) of conservation tillage farmers did. By comparison, 3.8 (0.6) percent of conventional tillage farmers used Bt-enhanced seed in 1995, while 2.4 (0.5) percent of conservation tillage farmers did. In both of these instances, differences are not statistically significant.

Machinery Use and Costs

Machinery-related costs typically range from $50 per acre per year to $70 per acre per year (Siemans and Doster, 1992) and overshadow all other cost categories except land. As with the choice of other inputs, a farmer endeavors to perform the requisite field operations with the optimum or least-cost machinery inventory. Because the optimum machinery inventory differs across tillage systems, direct comparison of machinery costs that can perform the desired field operations is difficult. As the size (width) of a machin-

TABLE 3.8. Average Seeding Rate by Tillage Practice—1990-1995

Commodity/Year	Conventional Tillage	Conservation Tillage
Corn	Kernels per acre	
1990	26,645 (84.847)[1]	24,613 (153.35)
1991	25,010 (102.62)	24,884 (172.47)
1992	25,521 (120.61)	25,542 (161.02)
1993	25,727 (69.084)	25,716 (90.012)
1994	26,002 (67.602)	25,809 (91.643)
1995	26,550 (109.88)	26,319 (128.87)
	Pounds per acre	
Soybeans		
1990	62.5 (0.42)	67.9 (1.05)
1991	62.4 (0.52)	64.9 (0.95)
1992	55.2 (0.87)	53.2 (1.16)
1993	66.8 (1.50)	72.9 (1.51)
1994	66.1 (1.46)	72.4 (1.65)
1995	55.1 (3.52)	62.8 (3.64)
Winter Wheat		
1990	74.1 (0.97)	69.2 (2.16)
1991	76.5 (0.98)	74.6 (2.93)
1992	75.3 (0.84)	69.2 (2.04)
1993	73.0 (0.86)	71.8 (2.40)
1994	71.9 (0.79)	74.8 (2.57)
1995	61.1 (1.13)	59.3 (2.32)
Spring Wheat		
1990	92.3 (1.34)	88.3 (2.45)
1991	89.0 (1.61)	84.4 (1.80)
1992	93.5 (1.52)	87.7 (1.72)
1993	89.3 (1.33)	94.0 (3.71)
1994	96.4 (1.84)	92.1 (2.22)
1995	94.2 (2.68)	92.1 (5.11)
Durum Wheat		
1990	97.10 (1.81)	97.83 (1.91)
1991	100.1 (1.67)	103.5 (2.21)
1992	93.82 (1.83)	97.55 (1.86)
1993	99.33 (1.33)	104.0 (8.71)
1994	99.70 (1.88)	103.3 (1.41)
1995	98.52 (3.24)	96.31 (2.93)

Source: USDA, Economic Research Service, *Cropping Practices Survey* (1990-1995, p. 51).

[1] Values in parentheses are the standard errors.

TABLE 3.9. Analysis of Variance of the Seeding Rate Based on the *Cropping Practices Survey*—1995

| Commodity | Variable | | |
	HEL[1]	CONTIL[2]	HEL *CONTIL[3]
Corn	2,7023.8 (ns)	2,548.8 (ns)	5,277.9 (ns)
Soybeans	15,425.9 (ns)	541.6 (ns)	6,092.2 (ns)
Winter Wheat	1,501.1 (ns)	50.6 (ns)	0.0 (ns)
Spring Wheat	17,304.1 (ns)	2,147.0 (ns)	1,660.5 (ns)
Durum Wheat	838.3 (ns)	286.3 (ns)	36.8 (ns)

Source: USDA, Economic Research Service, *Cropping Practices Survey* (1995, p. 64).

Note: The values in the table are the ANOVA sum of squares for the respective variables. The characters in parentheses indicate whether the sum of squares is not statistically significant (ns), whether it is statistically significant at the 5 percent level or better (*), or whether it is statistically significant at the 1 percent level or better (**).

[1]HEL indicates whether the land was designated as highly erodible land.
[2]CONTIL indicates whether conservation tillage was used.
[3]HEL*CONTIL is the interaction term.

ery set increases, machinery productivity increases. However, the annual machinery costs, fixed and variable, also increase with increasing machinery size. This is illustrated by comparing the farm machinery operating costs for planting equipment (see Table 3.10). For conservation tillage, fewer implements and field operations are used. If conservation tillage is used on only part of the cropland, however, implements and tractors will need to be available for other portions, so cost comparisons will be more difficult. Thus, using a drill or narrow-row planter for soybeans is an option for most tillage systems. However, owning a drill for soybeans and a planter for corn often increases the machinery inventory and costs for a corn-soybean farm.

Additionally, a farmer who decides to convert from conventional tillage to conservation tillage exclusively must consider how to value his or her existing conventional tillage equipment. This equipment might not be fully depreciated, and farmers often have limited alternatives on how it might otherwise be used (Bates, Rayner, and Custance, 1979; John Deere and Co., 1980). A complete assessment of the production costs associated with conservation tillage versus conventional tillage must make some provision for the opportunity cost of the conventional tillage equipment. No data are available, however, to permit a comparison of machinery costs between tillage systems (Beattie, Thompson, and Boehlje, 1974).

Machinery operating costs may not be lower for conservation tillage (see Table 3.10). Lower fuel and maintenance costs associated with conservation tillage may be overshadowed by the higher cost of new conservation tillage implements. Lower fuel costs are a consequence of the fewer trips across a field with conservation tillage. Maintenance costs will be lower because the equipment will be used less (Hunt, 1984).

Cost Comparison for Different Tillage Systems

As the literature and the ERS CPS data show, it is difficult to draw definitive conclusions regarding the aggregate effects on farm profitability of conservation tillage. As shown, profits are a function of many site-specific factors, including soil characteristics, local climatic conditions, cropping patterns, and other attributes of the overall farming operations. The decision on which tillage system to adopt must be made at the individual farm level and be based on many site-specific factors. Even in this situation, some inherent uncertainty exists. For example, depending on the weed problem in a specific year, the herbicide cost can be relatively larger or smaller. This can make a seemingly good ex ante decision to adopt conservation tillage a poor one ex post. Table 3.11 shows the results from field experiments with different tillage systems conducted in central Illinois (Siemans and Doster, 1992). The experiments were conducted coincidentally with the experiments on yields reported in Table 3.1. Along with machinery costs, herbicide use is the major input affected by a tillage system. As tillage is reduced, dependence

TABLE 3.10. Planting Equipment Operating Costs

Machine	Tractor size (hp)	Net cost of a new implement	Estimated work performed (acres per hour)	Total cost per hour	Total cost per acre[1]	Operating expense per acre	Diesel fuel (gallons per acre)
Row crop planter 8-30[2]	75	$20,290	9.33	$60.77	$6.51	$1.10	0.43
Row crop planter 12-30[2]	105	$32,637	14.00	$85.45	$6.10	$1.02	0.40
Minimum tillage planter 8-30	105	$27,640	8.48	$76.51	$9.02	$1.54	0.66
Minimum tillage planter 12-30	160	$46,599	12.73	$98.11	$7.71	$1.80	0.67
No tillage drill 30 ft	200	$52,423	12.73	$122.59	$9.63	$2.02	0.83

Source: Doane's Agricultural Report (1997, p. 3).

[1]Includes tractor, machinery, and labor costs.
[2]Used on conventional tillage systems.

on herbicides for weed control increases, but cost does not necessarily increase. In many field situations, as tillage is reduced, more expensive herbicide combinations and possibly a contact herbicide may be required to achieve adequate weed control. In many cases, herbicide cost is the same for conservation tillage and conventional tillage. Depending on herbicide cost, the production costs of conservation tillage can be greater than, equal to, or less than those for conventional tillage.

The cost comparisons from Table 3.11 are based on the assumption that the farmer is acquiring new equipment—either for conservation tillage or for conventional tillage. Changing tillage practices from conventional tillage to conservation tillage, however, requires careful consideration. Machinery purchases may be justified if soil erosion is a primary concern or if equipment purchase is part of a normal replacement schedule. Account must be taken of any remaining value of conventional tillage equipment in assessing the relative costs of conservation tillage versus conventional tillage.

ENVIRONMENTAL BENEFITS AND COSTS OF CONSERVATION TILLAGE

Background

A relatively small portion of cropland—that with high erosion rates—is responsible for a large proportion of total soil eroded in the United States (Magleby et al., 1995). Moreover, this cropland with high erosion rates is geographically concentrated. The soil erosion problem associated with agricultural production is not uniformly spread across the United States. Table 3.12 and Table 3.13 portray the geographic distribution of sheet and rill erosion in the United States as well as the distribution of wind erosion. In general, a soil erosion rate in excess of five tons per acre leads to a reduction in soil productivity (Alt, Osborn, and Colocicco, 1989). The greater the erosion rate, the greater the reduction in long-term agricultural productivity (the on-site cost of soil erosion) and impairment of water resources (the off-site cost of soil erosion).

TABLE 3.11. Machinery, Labor, and Herbicide Costs for Corn and Soybeans by Tillage System on a 1,000-Acre Corn-Soybean Farm in Central Illinois

After soybeans After corn	Chisel MB plow	Disk Chisel	Disk Disk	No tillage Chisel	No tillage No tillage	Ridge tillage Ridge tillage
			--- $/acre ---			
Machinery costs for corn	55	50	48	44	37	44
Labor[2]	8	7	7	7	5	7
Corn herbicides	10-15	10-15	10-15	15-25	15-25	5-25
Total	73-78	67-72	65-70	66-76	57-67	56-76
Expense (E) or Savings (S)[3]		1-11(S)	3-13(S)	3(E)-12(S)	6-12(S)	3(E)-22(S)
Costs for soybeans						
Machinery costs for soybeans	55	50	48	44	37	44
Labor	8	7	7	7	5	7
Soybean herbicides	14-28	14-28	14-28	14-28	25-40	7-40
Total	77-91	71-85	59-83	65-79	67-82	58-91
Expense (E) or Savings (S)[3]		8(E)-20(S)	5(E)-22(S)	2(E)-26(S)	5(E)-24(S)	14(E)-33(S)

Source: Siemans and Doster (1992, p. 49).

[1]Both corn and soybeans planted in rows. The least-cost machinery set is such that both corn and soybeans are planted and harvested in a timely manner so opportunity costs are negligible.
[2]Labor cost is assumed to be $8.50 per hour.
[3]Compared to chisel/moldboard (MB) plow system.

TABLE 3.12. Erosion and Crop Change for Conservation Tillage Adoption on HEL, State Level Results

STATE	ACRES (1000s)	USLE92 t/a/yr[1]	USLENEW t/a/yr	WEQ92 t/a/yr	WEQNEW t/a/yr	HAYCOV (1000s)	CONTIL (1000s)	NO CHANGE (1000s)
AL	1,288.3	6.6	4.5	0.0	0.0	160.6	532.7	595.0
AR	270.1	4.9	3.3	0.0	0.0	29.8	103.3	137.0
AZ	488.2	0.5	0.1	14.6	0.4	166.3	257.9	64.0
CA	879.7	4.1	3.4	3.2	0.6	284.9	250.7	344.1
CO	5,138.1	2.1	2.0	10.9	8.5	109.4	2,799.3	2,229.4
CT	23.0	4.0	2.1	0.0	0.0	3.6	3.9	15.5
DE	1.3	12.5	10.3	0.0	0.0	0.0	0.7	0.6
FL	167.5	3.4	2.6	0.0	0.0	71.3	46.4	49.8
GA	702.5	9.6	7.1	0.0	0.0	147.1	272.1	283.3
IA	3,980.5	6.7	6.3	0.5	0.4	38.7	1,592.9	2,348.9
ID	2,427.4	3.2	2.8	5.1	3.8	78.4	1,059.9	1,289.1
IL	3,098.8	10.2	7.6	0.0	0.0	36.6	2,245.6	816.6
IN	786.5	8.4	6.0	0.1	0.1	58.7	479.3	248.5
KS	7,265.0	2.3	2.0	2.8	2.5	8.6	4,753.9	2,502.5
KY	1,985.4	5.0	3.8	0.0	0.0	62.8	362.3	1,560.3
LA	138.3	3.0	2.9	0.0	0.0	16.0	69.0	53.3
MA	22.4	0.6	0.5	0.0	0.0	4.6	1.4	16.4
MD	229.8	9.1	6.3	0.0	0.0	10.8	110.4	108.7
ME	22.9	1.3	1.3	0.0	0.0	2.0	3.6	17.3
MI	472.9	5.8	4.1	1.1	0.8	50.6	198.9	223.4
MN	2,005.5	6.4	4.7	3.2	1.8	29.4	1,081.8	894.3
MO	3,965.6	7.6	5.7	0.0	0.0	104.8	1,391.6	2,469.2
MS	1,130.6	9.6	8.2	0.0	0.0	146.8	396.7	587.1
MT	10,041.3	1.7	1.2	6.5	3.0	4.4	6,430.1	3,606.8
NC	689.8	12.7	11.0	0.0	0.0	115.4	277.0	297.4
ND	6,001.8	1.8	1.3	3.1	0.9	98.7	3,308.0	2,595.1
NE	4,519.6	4.0	3.7	2.6	2.7	105.1	2,751.6	1,662.9
NH	26.0	1.2	1.1	0.0	0.0	0.0	1.1	24.9
NJ	54.8	6.3	4.3	0.1	0.1	10.7	15.4	28.7
NM	1,968.8	0.6	0.6	12.1	8.3	217.9	802.9	948.0
NV	365.8	0.1	0.0	25.8	5.0	130.7	0.9	234.2

STATE	ACRES (1000s)	USLE92 t/a/yr[1]	USLENEW t/a/yr	WEQ92 t/a/yr	WEQNEW t/a/yr	HAYCOV (1000s)	CONTIL (1000s)	NO CHANGE (1000s)
NY	1,581.2	4.1	3.2	0.0	0.0	81.9	363.7	1,135.6
OH	1,494.4	7.0	5.0	0.0	0.0	78.7	574.0	841.7
OK	2,532.2	2.6	2.2	3.0	3.0	14.2	1,508.6	1,009.4
OR	1,039.3	2.6	1.9	2.3	1.0	65.2	313.5	660.6
PA	3,520.0	4.4	4.3	0.0	0.0	190.5	1,392.2	1,937.3
RI	0.0	0.0	0.0	0.0	0.0	0.0	0.0	0.0
SC	327.6	5.4	3.3	0.0	0.0	66.5	128.1	133.0
SD	2,420.9	2.7	2.3	3.4	2.3	14.9	1,080.2	1325.8
TN	2,552.5	8.6	6.1	0.0	0.0	143.6	1,110.3	1298.6
TX	9,149.3	2.1	1.9	15.2	15.2	314.8	5,680.2	3,154.3
UT	551.6	0.8	1.0	7.3	5.6	82.7	97.0	371.9
VA	1,091.0	6.5	5.4	0.0	0.0	73.3	279.3	738.4
VT	121.3	2.5	1.6	0.0	0.0	0.0	12.7	108.6
WA	3,069.6	6.0	4.4	6.6	3.9	147.4	2,014.2	908.0
WI	2,723.6	5.9	5.1	0.0	0.0	58.2	1,137.4	1,528.0
WV	462.7	2.4	1.8	0.0	0.0	18.4	33.5	410.8
WY	1,264.2	1.1	1.1	14.1	13.8	7.7	465.5	791.0
US	94,059.7	4.1	3.2	4.6	4.0	3,662.7	47,791.7	42,605.3

Source: USDA, Natural Resources Conservation Service (1994, p. 61).

[1] t/a/yr denotes tons per acre per year.

ACRES is the HEL acreage that was not conservation tilled in 1992.
USLE92 is the value of the universal soil loss equation for 1992.
USLENEW is the value of the universal soil loss equation after changing to conservation tillage.
WEQ92 is the value of the wind erosion equation for 1992.
WEQNEW is the value of the wind erosion equation after changing to conservation tillage.
HAYCOV is acreage that was planted in any crop similar to hay cover.
NO CHANGE is acreage changing to a mix of crops with conversion to conservation tillage.
CONTIL is acreage changing to conservation tillage but the same crop is planted.

TABLE 3.13. Erosion and Crop Change for Conservation Tillage Adoption on HONHEL, State Level Results

STATE	ACRES (1000s)	USLE92 t/a/yr[1]	USLENEW t/a/yr	WEQ92 t/a/yr	WEQNEW t/a/yr	HAYCOV (1000s)	CONTIL (1000s)	NO CHANGE (1000s)
AL	2,325.6	4.7	3.6	0.0	0.0	359.0	1,326.1	640.5
AR	7,090.9	3.2	2.9	0.0	0.0	202.7	6,482.8	405.4
AZ	670.9	0.6	0.4	12.5	2.7	240.0	351.4	79.5
CA	7,981.7	0.4	0.3	0.4	0.2	2,441.0	3,576.9	1,963.8
CO	2,513.5	1.5	1.3	5.0	3.1	69.1	1,252.6	1,191.8
CT	205.5	3.3	1.2	0.0	0.0	44.0	56.6	104.9
DE	158.7	2.6	2.1	0.9	1.3	7.6	128.1	23.0
FL	2,768.5	1.0	0.7	0.0	0.0	1,310.4	852.0	606.1
GA	4,544.1	3.9	3.5	0.0	0.0	754.7	2,309.8	1,479.6
IA	7,695.8	2.6	2.2	1.8	1.1	163.5	6,246.0	1,286.3
ID	2,226.4	1.5	1.3	2.5	1.9	110.0	827.3	1,289.1
IL	14,386.7	3.0	2.2	0.0	0.0	66.9	13,554.6	765.2
IN	8,868.4	2.9	1.9	0.4	0.4	322.9	7,637.5	980.0
KS	14,339.0	2.2	1.9	1.3	1.2	15.5	12,276.2	2,047.3
KY	1,762.1	3.2	2.1	0.0	0.0	28.2	946.3	787.6
LA	5,800.7	3.5	3.3	0.0	0.0	933.4	4,004.6	862.7
MA	247.0	1.3	1.4	0.0	0.0	42.3	38.4	166.3
MD	505.8	3.6	3.1	0.1	0.1	32.2	342.4	131.2
ME	421.9	1.1	1.1	0.0	0.0	21.9	34.6	365.4
MI	7,149.2	1.5	1.2	2.2	1.4	620.0	4,088.9	2,440.3
MN	18,945.5	1.5	1.1	5.6	3.9	446.3	14,192.1	4,307.1
MO	8,639.0	3.3	2.3	0.0	0.0	256.6	6,668.0	1,714.4
MS	4,772.0	3.9	3.7	0.0	0.0	558.7	3,738.8	474.5
MT	4,095.2	1.1	0.9	3.1	1.5	7.3	2,127.8	1,960.1
NC	4,921.8	4.1	3.0	0.0	0.0	484.8	3,338.3	1,098.7
ND	20,561.0	1.1	0.8	1.4	0.6	504.7	16,746.7	3,309.6
NE	11,314.5	2.9	2.6	1.0	0.9	319.4	9,170.1	1,825.0
NH	114.6	0.8	0.9	0.0	0.0	7.7	14.1	92.8
NJ	499.8	4.2	3.4	0.0	0.0	111.0	237.4	151.4
NM	57.0	0.3	0.3	3.9	3.1	7.3	12.8	36.9
NV	396.5	0.0	0.0	0.7	0.1	27.3	4.9	364.3

STATE	ACRES (1000s)	USLE92 t/a/yr[1]	USLENEW t/a/yr	WEQ92 t/a/yr	WEQNEW t/a/yr	HAYCOV (1000s)	CONTIL (1000s)	NO CHANGE (1000s)
NY	3,990.3	1.8	1.7	0.0	0.0	269.3	1,305.8	2,415.2
OH	7,640.6	2.6	1.5	0.1	0.1	230.5	6,068.2	1,341.9
OK	4,467.3	3.2	2.6	0.8	0.8	84.6	3,787.2	595.5
OR	2,321.3	2.0	1.5	0.5	0.6	155.0	1,168.9	997.4
PA	1,342.3	2.8	2.6	0.0	0.0	57.4	568.0	716.9
RI	24.6	2.5	0.9	0.0	0.0	7.2	3.3	14.1
SC	2,754.3	2.8	2.4	0.0	0.0	237.7	2,013.4	503.2
SD	13,217.5	1.7	1.4	1.9	1.4	152.4	9,890.7	3,174.4
TN	2,037.9	3.9	2.5	0.0	0.0	77.2	1,430.0	530.7
TX	13,014.8	2.6	2.4	3.6	2.4	1,126.8	9,721.7	2,166.3
UT	1,375.0	1.0	1.0	2.4	1.6	68.5	473.3	833.2
VA	1,460.5	2.8	2.5	0.2	0.2	69.1	711.1	680.3
VT	503.5	1.0	1.0	0.0	0.0	5.8	94.2	403.5
WA	4,104.0	2.1	1.7	2.8	1.5	271.0	2,603.8	1,229.2
WI	7,409.6	1.9	1.7	0.2	0.3	122.2	3,823.3	3,464.1
WV	387.2	1.0	0.6	0.0	0.0	17.9	64.0	305.3
WY	1,064.6	0.3	0.2	1.9	1.9	12.8	144.3	907.5
US	233,094.6	2.3	1.8	1.4	1.0	13,481.8	166,455.3	53,157.5

Source: USDA, Natural Resources Conservation Service (1994, p. 67).

[1] t/a/yr denotes tons per acre per year.

ACRES is the HEL acreage that was not conservation tilled in 1992.
USLE92 is the value of the universal soil loss equation for 1992.
USLENEW is the value of the universal soil loss equation after changing to conservation tillage.
WEQ92 is the value of the wind erosion equation for 1992.
WEQNEW is the value of the wind erosion equation after changing to conservation tillage.
HAYCOV is acreage that was planted in any crop similar to hay cover.
NO CHANGE is acreage changing to a mix of crops with conversion to conservation tillage.
CONTIL is acreage changing to conservation tillage but the same crop is planted.

The most widespread off-site erosion-related problem is impairment of water resource use (National Research Council, 1993). The U.S. Environmental Protection Agency has identified siltation associated with erosion in rivers and lakes as the second leading cause of water quality impairment, and agricultural production is identified as the leading cause of water quality impairment (U.S. Environmental Protection Agency, 1995).

Three related causes of water use impairment are sedimentation, eutrophication, and pesticide contamination. When soil particles and agricultural chemicals wash off a field, they may be carried in runoff until discharged into a water body or stream. Not all agricultural constituents that are transported from a field reach water systems, but a significant portion does, especially dissolved chemicals and the more chemically active, finer soil particles. Once agricultural pollutants enter a water system, they lower water quality and can impose economic losses on water users. These off-site impacts can be substantial. The off-site impacts of erosion are potentially greater than the on-site productivity effects in the aggregate (Foster and Dabney, 1995). Therefore, society may have a larger incentive for reducing erosion than farmers have.

If the runoff reaches the water body or stream, soil particles can be suspended in the water or settle out as sediment, depending on the velocity of the water flow and the size of the soil particles. In each case, water use can be affected.

Suspended soil particles affect the biologic nature of water systems by reducing the transmission of sunlight, raising surface water temperatures, and affecting the respiration and digestion of aquatic life. The effects on aquatic life and the reduction in aesthetic quality of recreation sites can reduce the value of water for recreation uses. Suspended soil particles impose costs on water treatment facilities that must filter out the particles. Suspended soil particles can also damage moving parts in pumps and turbines.

Even when soil particles settle on the bottom of a river or lake, they can cause serious problems for aquatic life by covering food sources, hiding places, and nesting sites. Sedimentation can clog navigation and water conveyance systems such as roadside ditches, reduce reservoir capacity, and damage recreation sites. In streambeds, sedimentation can lead to an increase in the frequency and severity of flooding by reducing channel capacity.

The nutrients and pesticides attached to soil particles, or dissolved in runoff, affect water quality in ways that can affect the suitability of water for many uses (D. Baker, 1987). The most far-reaching impact is eutrophication, abundant growth of algae and rooted vegetation caused by excessive nutrient runoff. As algae die and decay, they use oxygen from the surrounding water, lowering the dissolved oxygen levels and altering the size and composition of commercial and recreational sport fisheries. Rooted plants can become a nuisance around marinas and shorelines. Floating algae blooms can restrict light penetration to surface waters and can affect the health, safety, and enjoyment of people using water for recreation. Floating algae can clog intake pipes and filtration systems, increasing the cost of water treatment.

Pesticides, which include herbicides, insecticides, and fungicides, create a broad array of impacts. Most notable are effects on aquatic life. Very high concentrations will kill organisms outright. Lower concentrations, more commonly observed, can produce a variety of sublethal effects such as to lower resistance of fish, which makes them susceptible to other stresses (Glotfelty, 1987). Herbicides can hinder photosynthesis in aquatic plants (Schepers, 1987). Pesticides can damage commercial and sport fisheries and make fish dangerous to eat (Herndon, 1987).

Several studies are available that evaluate the improvements in water quality associated with the use of conservation tillage. A few of these are reviewed here. Richards and Baker (1998) report on the effort to reduce the eutrophication in Lake Erie that began in the early 1970s. A monitoring station was set up at Bowling Green, Ohio, on the Maumee River, which feeds into Lake Erie. Agriculture is the dominant land use in the Maumee River basin. Major crops are corn, soybeans, and wheat. Between 1975 and 1995, implementation of no tillage and reduced tillage (see Figure 2.1) increased from less than 5 percent to more than 50 percent of planted acreage. Fertilizer (nitrogen and phosphorous) application rates also changed over the period so it is not possible to quantify precisely the contribution of conservation tillage to the water quality improvements. Nevertheless, water quality changes over the study period were evaluated by conducting trend studies of concentrations and loads. The adoption of conservation tillage in conjunction with the reduced fertilizer application rates led to a reduction in total phosphorous loadings of 24 percent, a reduction in suspended sediments of 19 percent, and a reduction in total Kjeldahl nitrogen of 10 percent.

Fawcett and colleagues (1994)[12] survey the effects of various best management practices, including conservation tillage, on pesticide runoff into surface water and leaching into groundwater. They conclude that no tillage systems provide a reduction in runoff losses for active pesticide ingredients studied. Average[13] herbicide runoff in no tillage systems, for example, was 30 percent of the conventional tillage runoff. Also, ridge tillage and chisel plow practices are less effective than no tillage in reducing soil erosion on HEL, but are relatively good production practices on less erodible fields. For the various conservation tillage practices relative to conventional tillage, herbicide runoff was 70 percent less for no tillage, while it was 42 percent less for ridge tillage. With regard to leaching, however, conservation tillage does not fare as well. Increases in infiltration accompanying the use of conservation tillage may result in a greater threat to groundwater from pesticides or nitrate. Preferential flow of water through macropores, which may be more prevalent with no tillage, can allow water and dissolved solutes or suspended sediment to bypass upper layers of soil. This may transfer pesticides to shallow groundwater or to depths in the soil where biological degradation is slower. It is important, however, to keep this in perspective. Even though an increase in the potential leaching risks of certain pesticides is associated with conservation tillage, the relative concentrations of pesticides found in surface water are typically greater than concentrations in groundwater.

Wind erosion produces off-site impacts that can be as dramatic as the Dust Bowl of the 1930s. It has not, however, received the attention given to the more widespread water erosion impacts. Damage can include higher maintenance of buildings and landscaping, pitting of automobile finishes and glass, greater wear on machinery parts, increased soiling and deterioration of retail inventories, costs of removing blown sand and dust from roads and ditches, and increased respiratory and eye disorders. Off-site damages from wind erosion depend on the extent and location of population centers relative to prevailing winds and wind erosion sources (Piper and Lee, 1989). Consequently, damage estimates for one area cannot readily be extrapolated to other areas, nor can the impact of wind erosion from cropland or other agricultural land be differentiated from wind erosion originating on nonagricultural land.

Off-site impacts of both sheet and rill erosion and wind erosion may be subject to threshold effects (Zison, Haven, and Wills, 1977). A reduction in erosion may not produce proportional improvements in water or air quality unless they are quite large in relation to total loads. In economic terms, the costs of erosion control practices that result in only small reductions in erosion may produce few, if any, off-site benefits.

A third and somewhat ancillary erosion-related problem deals with wildlife. Monoculture production and field consolidation have diminished habit diversity in areas where agriculture once contributed to diversity (Strohbehn, 1986). Soil conservation practices frequently enhance wildlife habitat. Field borders, windbreaks, hedgerows, stream bank protection, and wildlife habitat management can increase habitat diversity. Practices aimed at wildlife protection, however, often divert land from row crop production, thereby creating opportunity costs.

Analysis of Potential Soil Erosion Reduction from Conservation Tillage Adoption

Measuring the Off-Site Benefits of Soil Erosion Reduction

To quantify the benefits of reduced soil erosion, from both water and wind, it is necessary first to determine the reduction in soil erosion associated with conservation tillage relative to conventional tillage and then calculate the social benefits associated with this reduction. An approach that has been frequently used to measure soil losses by sheet and rill erosion is the universal soil loss equation (USLE) (Wischmeier and Smith, 1978).[14] Wind-related soil losses are measured by the wind erosion equation (WEQ) (Skidmore and Woodruff, 1980; Smith and English, 1982). These are the relationships currently used by the Natural Resources Conservation Service of the U.S. Department of Agriculture to calculate the soil loss on cropland in the United States. They are an integral part of the conservation practice standards used in determining conservation compliance (USDA, Natural Resources Conservation Service, 1997a). They will be used here to assess the reduction in soil erosion associated with an increase in the use of conservation tillage. The appendix at the end of this chapter provides a technical definition of each type of soil loss. It also provides a technical definition of highly erodible land.[15]

Before using the USLE and WEQ to assess the benefits of increasing the use of conservation tillage, a discussion of the data used in the analysis is needed.[16]

National Resources Inventory

Every five years since 1977, the Natural Resources Conservation Service (NRCS) has conducted a statistically representative National Resources Inventory (NRI) of land cover and use, soil erosion, prime farmland, and other natural resource statistics on non-Federal, rural land. Based on actual field observations by NRCS technicians, the NRI provides a profile of the nation's conservation practices and future program needs. The most recent complete inventory was conducted in 1992 (USDA, Soil Conservation Service, 1992).[17] The data consist of information on 236 variables collected at approximately 1 million sites, with most cropland points representing 1,500 to 2,600 acres. Information on all of the components of the USLE and WEQ is reported.

The NRI data show that 94 million acres of highly erodible land (HEL) were not conservation tilled in 1992, of which 24.5 million acres had USLE erosion rates above the soil loss tolerance level (T) and 22.2 million acres had wind erosion rates above T.[18] Of this HEL, about 15 million acres had either or both USLE or wind erosion rates above 2*T, which implies that continued production would not be sustainable. In addition, the data show that for NON-HEL cropland only about 10 percent (20 million acres) had either USLE or wind erosion rates or both above the soil loss tolerance level.

Implications of Reducing Soil Erosion by Converting
All Cropland to Conservation Tillage

An evaluation was made of the implications of reducing soil erosion by converting all cropland to conservation tillage based on the 1992 NRI data. Table 3.14 defines for each crop or land use the assumption made about the adoption of conservation tillage. The cropland uses recorded in the NRI were divided into three groups, each of which required a different set of procedures. The first group of crops used for this analysis are those for which conservation tillage is either not practical or it is technologically infeasible. This group includes intensively tilled crops, such as potatoes and vegetables, and hay crops for which tillage is already very limited. For these crops, tillage type and erosion rates are assumed to remain unchanged. The acres of this group are shown in the column labeled NO CHANGE in Tables 3.12 and 3.13, while the erosion rates for each crop can be seen in Tables 3.15 and 3.16. This group constitutes 43 million acres of HEL and 53 million acres of NONHEL cropland.

The second cropland use category in this analysis is the group of major field crops for which high rates of conservation tillage adoption already exist. For these crops, it is assumed that acres currently not conservation tilled could switch. For sheet and rill erosion, it is assumed that for each crop, a switch to conservation tillage would change the USLE residue management factor (C) to that where the crop is conservation tilled.[19] Since sufficient information is not available for recalculating the wind erosion equation, it is assumed that for each crop, a switch to conservation tillage will result in the same wind erosion rate as crops currently conservation tilled. The acres in this group are shown in the column labeled CONTIL in Tables 3.12 and 3.13, while erosion rates for each crop are presented in Tables 3.15 and 3.16. This group constitutes 48 million acres of HEL and 166 million acres of NONHEL cropland.

The third group of crops are mainly fruit, nuts, and Conservation Reserve Program (CRP) acreage for which it is assumed that a grass cover will be maintained, resulting in the same erosion rates as a hay crop. The number of acres of crops in this group are in the column labeled HAYCOV in Tables 3.12 and 3.13, while erosion rates are shown in Tables 3.15 and 3.16. This group is relatively small, having 3.6 million acres of HEL and 13.5 million acres of NONHEL.

The results of the analysis show that the total erosion reduction due to the assumed adoption of conservation tillage is 326 million

TABLE 3.14. Crop Management Assumptions for Changing to Conservation Tillage

Land Use	No Change[1]	Switch to Conservation Tillage[2]	Cover Management Switch[3]
Fruit			X
Nuts			X
Vineyard			X
Bush fruit			X
Berries			X
Horticulture – other			X
Corn		X	
Sorghum		X	
Soybeans		X	
Cotton		X	
Peanuts	X		
Tobacco	X		
Sugar beets	X		
Potatoes	X		
Vegetables – other	X		
Row crops – other	X		
Sunflowers		X	
Wheat		X	
Oats		X	
Rice		X	
Barley		X	
Close – other		X	
Hay/grass	X		
Hay/legume	X		
Hay/legume/grass	X		
Summer fallow	X		
Not planted			X
CRP[4]	X		

Source: USDA, Natural Resources Conservation Service (1994, p. 35).

[1] Erosion of these crops is already equal to that of conservation tillage or else the crop is very unlikely to switch to conservation tillage.

[2] It is assumed these crops switch to conservation tillage. The contribution to soil loss is assumed to be similar to other planted acreage in the area that has already switched.

[3] The crop cover is assumed to be similar to a hay crop and to contribute to soil loss in a way similar to hay.

[4] Conservation Reserve Program acreage.

TABLE 3.15. Erosion and Crop Change for Conservation Tillage Adoption on HEL, Totals by Crop

Land Use	ACRES (1000s)	USLE92 [t/a/yr]¹	USLENEW t/a/yr	WEQ92 t/a/yr	WEQNEW t/a/yr
Fruit	638.9	1.7	1.7	0.7	1.4
Nuts	136.2	1.4	0.5	2.6	1.8
Vineyard	102.0	2.6	1.6	0.3	1.8
Bush fruit	0.9	0.5	0.5	0.0	0.0
Berries	18.9	2.9	1.9	0.0	0.0
Horticulture-other	143.4	6.4	1.5	1.3	0.4
Corm	10,444.6	10.4	8.2	2.8	2.0
Sorghum	3,647.2	3.6	3.1	13.0	10.1
Soybeans	4,566.8	14.1	10.8	0.7	0.5
Cotton	3,339.5	6.0	5.3	22.1	22.1
Peanuts	411.1	8.5	8.5	9.3	9.3
Tobacco	366.5	17.1	17.1	0.0	0.0
Sugar beets	269.1	1.7	1.7	12.4	12.4
Potatoes	393.6	2.1	2.1	15.3	15.3
Vegetables—other	613.1	6.5	6.5	9.1	9.1
Row crops—other	505.7	2.4	2.4	8.4	8.4
Sunflowers	113.8	1.4	0.6	6.6	2.3
Wheat	14,434.0	3.7	2.9	7.4	4.5
Oats	1,134.8	5.5	4.9	2.5	1.0
Rice	20.8	3.8	0.0	0.0	0.0
Barley	1,509.4	5.1	4.0	6.9	3.2
Close—other	1,032.8	4.7	3.6	5.0	2.4
Hay/grass	7,752.7	1.1	1.1	0.2	0.2
Hay/legume	4,898.2	1.3	1.3	2.4	2.4
Hay/legume/grass	7,597.1	2.0	2.0	0.2	0.2
Summer fallow	7,548.0	3.4	2.7	9.0	5.5
Not planted	2,622.4	4.4	1.2	10.4	1.3
CRP	19,796.2	0.8	0.8	1.1	1.1
US	94,059.7	4.1	3.2	4.6	4.0

Source: USDA, Natural Resources Conservation Service (1994).

¹t/a/yr denotes tons per acre per year.

See Table 3.12 for a definition of column headings.

tons per year (see Tables 3.12 and 3.13). For HEL, the reduction in sheet and rill soil erosion rates from 4.1 to 3.2 tons per acre per year and the reduction in wind-related erosion rates from 4.6 to 4.0 tons per acre per year result in an average savings of 1.3 tons per acre per year on 94 million acres. For NONHEL, the reduction in sheet and rill soil erosion rates from 2.3 to 1.8 tons per acre per year and the reduction in wind-related soil erosion rates from 1.4 to 1.0 tons per

TABLE 3.16. Erosion and Crop Change for Conservation Tillage Adoption on NONHEL, Totals by Crop

Land Use	ACRES (1000s)	USLE92 ^{r}t/a/y^{r1}	USLENEW t/a/y^r	WEQ92 t/a/y^r	WEQNEW t/a/y^r
Fruit	2,195.9	0.6	0.2	0.1	0.1
Nuts	1,027.9	0.4	0.2	0.0	0.2
Vineyard	517.8	0.7	0.2	1.0	0.2
Bush fruit	55.3	0.7	0.2	0.2	0.2
Berries	167.3	1.0	0.3	0.1	0.0
Horticulture—other	429.8	2.0	0.3	0.4	0.1
Corn	48,040.8	3.0	2.4	1.4	0.9
Sorghum	7,105.1	3.2	3.0	2.7	1.8
Soybeans	39,109.4	3.4	2.6	1.1	0.7
Cotton	9,547.4	4.1	4.0	2.5	1.0
Peanuts	1,547.6	5.5	5.5	1.0	1.0
Tobacco	863.9	5.8	5.8	0.0	0.0
Sugar beets	958.9	1.0	1.0	8.0	8.0
Potatoes	750.8	2.1	2.1	3.4	3.4
Vegetables—other	2,231.3	2.6	2.6	1.6	1.6
Row crops—other	1,965.3	2.0	2.0	2.3	2.3
Sunflowers	1,575.7	1.3	1.0	3.3	2.3
Wheat	38,070.8	2.1	1.7	2.2	1.6
Oats	2,916.3	2.2	2.0	1.4	1.0
Rice	3,626.0	2.1	2.0	0.0	0.0
Barley	4,175.8	1.3	1.0	2.8	1.9
Close—other	3,232.7	1.6	1.5	1.3	0.6
Hay/grass	10,690.3	0.4	0.4	0.1	0.1
Hay/legume	8,222.7	0.6	0.6	0.6	0.6
Hay/legume/grass	11,627.7	0.7	0.7	0.2	0.2
Summer fallow	9,,055.3	2.1	1.8	3.1	2.0
Not planted	9,087.8	1.7	0.4	1.5	0.3
CRP	14,243.8	0.4	0.4	0.3	0.3
US	233,094.6	2.3	1.8	1.4	1.0

Source: USDA, Natural Resources Conservation Service (1994).

[1]t/a/yr denotes tons per acre per year.

See Table 3.12 for a definition of column headings.

acre per year result in a savings of 0.9 ton per acre per year on 233 million acres. There is variability in the erosion rates geographically and by crop. Some of this variability is an artifact of the assumptions made, and some is a function of the erosion potential of the cropland. Thus, for example, corn and soybeans, which are planted in regions where highly erodible land is relatively common, have the potential for realizing substantial reductions in the erosion rate, whereas peanuts and tobacco have no potential for contributing to a reduction in the erosion rate (see Table 3.15).

Social (Off-Site) Benefits of Converting
Highly Erodible Cropland to Conservation Tillage

The change in erosion rates is one method of assessing the benefits of switching to conservation tillage. The benefit measure, however, is in physical units (tons per acre per year). Such a measure does nothing to quantify the off-site social benefits of a reduction in soil erosion. This requires estimates of the off-site damages associated with erosion.

Ribaudo (1989) developed comprehensive estimates of the off-site damages associated with sheet and rill erosion.[20] The approach has been applied in a number of settings (e.g., Magleby et al., 1995; Ribaudo et al., 1990). The estimates take into account damage to water uses such as recreation, water storage facilities, commercial fishing, navigation, water storage, drinking water supplies, industrial water supplies, and irrigation. The estimates are compiled from an eclectic assortment of studies. These estimates will be used here with the assumption that any reduction in off-site damages translates into a comparable increase in social benefits.

Huzsar and Piper (1986) have derived estimates of the off-site damages due to wind erosion. There is considerable uncertainty, however, in quantifying the damages due to wind erosion. This uncertainty is a function of the poor understanding of households' response to blowing soil and how damages vary with population density. Any better alternative being absent, however, these estimates will be used with the assumption that any reduction in off-site damages will lead to a comparable increase in social benefits.

Combining the estimates of sheet and rill and wind erosion damages with data on the amount of HEL that is not treated with an acceptable conservation system and with the previously presented results on changes in the erosion rate if HEL is converted to conservation tillage, it is possible to estimate the social benefits of bringing the remaining HEL under conservation tillage. This, of course, presumes that the untreated HEL is switched to conservation tillage. Note that only untreated HEL is used in the analysis because it has been the focus of conservation compliance. In 1996, 22.4 million HEL acres in the United States were not adequately treated using some type of conservation management system or technical practice. Additional social benefits will result if conventional-tilled NONHEL is switched to conservation tillage, but the per acre benefits would be lower. A comparable number of acres of NONHEL switched, however, will result in a smaller social benefit because the erosion rate on NONHEL is lower than it is for HEL (see Tables 3.12 and 3.13).

Data on the number of untreated HEL acres for 1996 are taken from the Conservation Technology Information Center (1996). Table 3.17 presents the social benefits by state associated with a reduction in sheet and rill erosion, and Table 3.18 presents social benefits for a reduction in wind erosion. Wind erosion estimates are calculated only for states in four regions, Mountain, Northern Plains, Pacific, and Southern Plains. The soil erosion estimates derived are applicable only to states in these regions (Piper, 1990).

The best (most likely to be realized) estimate for the social benefits of a reduction in sheet and rill soil erosion is $32 million annually. For wind erosion, the estimate is $17.6 million annually. These estimates are likely overstated because it is not feasible to convert all HEL to conservation tillage. Some soils must be tilled regardless of whether they are HEL.[21] This can involve conventional tillage practices or mulch tillage or ridge tillage.

The Realized Social (Off-Site) Benefits of Conservation Tillage in 1996

Using the same approach as that used for computing the social benefits of converting the remaining highly erodible cropland to conservation tillage, it is possible to estimate the change in soil erosion and the social benefits associated with the use of conservation tillage in 1996. By assuming that all conservation-tilled acres in 1996 had been conventionally tilled and comparing the associated erosion rates to the erosion rates for the same acres conservation tilled, estimates of both the physical reduction in soil loss and the social benefits can be made. Data on the number of acres that were conservation tilled in 1996 were taken from the Conservation Technology Information Center (1996). An estimate of the number of conservation-tilled acres that are highly erodible is based on the percent of total cropland acres in a state that are highly erodible. The estimated changes in sheet and rill and wind erosion rates

TABLE 3.17. Annual Benefits of Conservation Tillage Associated with Reduced Sheet and Rill Erosion on Untreated HEL (in Millions of Dollars)

State	Range	Best[1]
AL	0.964861-2.234850	1.5833.62
AR	0.025986-0.142659	0.042554
AZ	0	0
CA	0	0
CO	0.15727-0.4268770	0.279592
CT	0.066952-0.224232	0.112275
DE	0.004066-0.013618	0.006819
FL	0.017891-0.041439	0.029359
GA	0.127612-.0.295580	0.209414
IA	1.061279-3.779644	2.141178
ID	0.279167-0.757739	0.496297
IL	1.135448-4.043787	2.290815
IN	1.298028-4.622802	2.618829
KS	0.069799-0.315343	0.071046
KY	0.552565-1.593937	0.998867
LA	0.003328-0.018270	0.005450
MA	0.000527-0.001765	0.000884
MD	0.929849-3.114222	1.559320
ME	0	0
MI	0.12904-0.3871210	0.241305
MN	1.017069-3.051208	1.901919
MO	1.218118-4.338211	2.457607
MS	0.232403-1.275879	0.380580
MT	0.100824-0.273666	0.179243
NC	0.616688-1.778908	1.114782
ND	0.083497-0377227	0.084988
NE	0.117632-0.531445	0.119733
NH	0.00038-0.0012720	0.000637
NJ	0.17645-0.5909590	0.295899
NM	0	0
NV	0.00393 0.0106660	0.006986
NY	0.828075-2.773361	1.388648
OH	0.210659-0.750243	0.425015
OK	0.088441-0.299160	0.155348
OR	0.097773-0.304182	0.158481
PA	0.761491-2.550363	1.276990
RI	0	0
SC	0.077263-0.178959	0.126790
SD	0.018997-0.085824	0.019336
TN	0.361343-1.042335	0.653197
TX	0.351075-1.187550	0.616671
UT	0	0
VA	0.601891-1.736225	1.088034
WA	1.287724-4.006252	2.087291
WI	1.777824-5.333472	3.324531
WV	0.003831-0.011050	0.006925
WY	0	0
US	18.28702-64.00455	31.91263

[1]Benefits deemed most likely.

TABLE 3.18. Annual Benefits of Conservation Tillage Associated with Reduced Wind Erosion on Untreated HEL (in Millions of Dollars)

State	Range	Best[1]
AZ	0	0
CA	0	0
CO	4.493441-12.98105	6.490526
ID	1.994049-5.760586	2.880293
KS	0.457019-1.329510	0.623208
MT	0.576138-1.664400	0.832200
ND	0.328023-0.954249	0.447304
NE	0.770210-2.240611	1.050287
NM	0.107585-0.310798	0.155399
NV	0.112276-0.324355	0.162177
OK	0	0
OR	0.082161-0.246485	0.100420
SD	0.000088-0.000256	0.000120
TX	0	0
UT	0.016682-0.48193	0.024096
WA	0.473427-1.420284	0.578634
WY	0.207288-0.598832	0.299416
US	12.19633-35.09782	17.59693

[1]Benefits deemed most likely.

associated with a change in tillage practice are those given in Tables 3.12 and 3.13. Estimates of off-site damages associated with sheet and rill erosion are taken from Ribaudo (1989), while estimates of off-site wind erosion damage comes from Huzsar and Piper (1986).

Table 3.19 presents state-level estimates of the reduction in soil erosion associated with using conservation tillage instead of conventional tillage. The table also presents the social benefits associated with a reduction in sheet and rill erosion. Table 3.20 presents erosion reduction and social benefits associated with wind erosion. As before, wind erosion estimates are calculated only for states in four regions, Mountain, Northern Plains, Pacific, and Southern Plains, because the soil erosion estimates derived are applicable only to states in these regions.

Sheet and rill erosion was reduced by about 66 million tons in 1996 in the United States because of conservation tillage (see Table 3.19). The best estimate of social benefits of the reduction is $103 million. For wind erosion, the estimate of the reduction in soil erosion is

TABLE 3.19. Benefits of Conservation Tillage in 1996 Associated with Reduced Sheet and Rill Erosion

State	Soil Erosion (million tons per year)	Nominal Soil Erosion Benefits (million $)	
		Range	Best[1]
AL	0.864994	1.012043-2.344133	1.660788
AR	0.399424	0.595142-3.267292	0.974596
AZ	0.01102	0.006943-0.018845	0.012343
CA	0.16991	0.259962-0.808771	0.421377
CO	0.166002	0.104582-0.283864	0.185923
CT	0.020857	0.087809-0.294087	0.147252
DE	0.191515	0.80628-2.7003670	1.352098
FL	0.046662	0.054595-0.126454	0.089591
GA	0.337876	0.395315-0.915643	0.648722
IA	4.659493	2.655911-9.458770	5.358417
ID	0.401726	0.253087-0.686951	0.449933
IL	10.61553	6.050853-21.54953	12.20786
IN	6.553143	3.735292-13.30288	7.536114
KS	1.949699	1.091832-4.932739	1.111329
KY	3.078445	2.401187-6.926502	4.340608
LA	0.174027	0.2593-1.42354101	0.424626
MA	0.000052	0.000218-0.000731	0.000366
MD	1.084735	4.566734-15.29476	7.658228
ME	0	0	0
MI	1.403866	2.807731-8.423194	5.250457
MN	2.880998	5.761997-17.28599	10.77493
MO	6.837683	3.897479-13.88050	7.863336
MS	0.592648	0.883046-4.847862	1.446062
MT	1.544968	0.97333-2.6418950	1.730364
NC	1.253669	0.977862-2.820755	1.767673
ND	2.128536	1.19198-5.3851950	1.213265
NE	2.743571	1.5364-6.94123400	1.563835
NH	0.000011	0.000046-0.000153	0.000077
NJ	0.15354	0.646403-2.164914	1.083992
NM	0	0	0
NV	0.000137	0.000087-0.000235	0.000154
NY	0.140663	0.59219-1.9833440	0.993079
OH	5.782275	3.295897-11.73802	6.649617
OK	0.898445	1.033211-3.494950	1.814858
OR	0.27649	0.42303-1.3160920	0.685695
PA	0.158062	0.665441-2.228674	1.115918
RI	0.000375	0.001577-0.005282	0.002645
SC	0.195293	0.228493-0.529245	0.374963
SD	1.646437	0.922005-4.165487	0.938469
TN	3.358827	2.619885-7.557361	4.735947
TX	0.997953	1.147646-3.882037	2.015865
UT	0	0	0
VA	0.488264	0.380846-1.098593	0.688452
VT	0.000695	0.002925-0.009798	0.004906
WA	0.897518	1.373202-4.272184	2.225844
WI	0.909727	1.819455-5.458364	3.402380
WV	0.027981	0.021825-0.062958	0.039454
WY	0.005005	0.003153-0.008559	0.005606
US	66.04875	57.54423-196.5387	102.9680

[1]Benefits deemed most likely.

TABLE 3.20. Benefits of Conservation Tillage in 1996 Associated with Reduced Wind Erosion

State	Soil Erosion (million tons per year)	Nominal Soil Erosion Benefits (million $) Range	Best[1]
AZ	0.53380	0.960833-2.775739	1.387869
CA	0.45504	0.409536-1.228607	0.500544
CO	2.96918	2.969182-7.719873	3.859936
ID	1.27875	2.301752-6.649507	3.324753
KS	1.21778	1.33956-3.8969030	1.826673
MT	6.78978	6.789784-17.65344	8.826720
ND	7.11476	3.557381-11.38362	4.980333
NE	0.45616	0.501781-1.459726	0.684247
NM	1.34726	2.425067-7.005749	3.502874
NV	0.03116	0.056095-0.162051	0.081025
OK	0	0	0
OR	0.21175	0.190576-0.571727	0.232926
SD	3.22511	3.547622-10.32036	4.837667
TX	3.97285	5.959276-16.68597	9.137557
UT	0.05297	0.095341-0.275429	0.137715
WA	1.81855	1.636694-4.910082	2.000404
WY	0.01084	0.019507-0.056355	0.028177
US	31.48575	32.75999-92.75513	45.34942

[1]Benefits deemed most likely.

31.5 million tons (Table 3.20). The best estimate of the social benefits associated with the reduction is $45 million.

The Realized Social (Off-Site) Benefits of Conservation Compliance

A significant change in U.S. conservation policy came in the Food Security Act of 1985 in the form of conservation compliance.[22] Although meeting the conservation provisions remains voluntary, a farmer who wants to receive certain agricultural program payments and whose cropland is designated as HEL has no choice but to implement an acceptable conservation plan. Requirements for conservation compliance were applied to HEL previously cultivated in any year between 1981 and 1985. Conservation compliance required farmers producing crops on HEL to implement and maintain an approved soil conservation system by 1995.

In 1996, the Natural Resources Conservation Service conducted a status review of conservation compliance through 1995 to determine its effect on aggregate soil loss (USDA, Natural Resources Conservation Service, 1996). The estimated changes are categorized by crop residue cover and reported in physical units.[23] The estimates are reproduced in Table 3.21. Using the same approach as that used for computing the social benefits of converting the remaining highly erodible cropland to conservation tillage, it is possible to provide a nominal estimate of the social benefits associated with conservation compliance through 1995.

As a result of conservation compliance, it is estimated that sheet and rill erosion, on average, has been reduced by about 61 percent on HEL in the United States. This translates into a best estimate of realized social benefits of the reduction of $579 million (see Table 3.21).

The Impact of Conservation Tillage on Wildlife and Wildlife Habitat

The use of a particular tillage system has several effects on wildlife. The amount of crop residue cover on the soil surface differs depending on the tillage practice used. Other factors include the availability of waste grain and other wildlife foods, the frequency and extent of disturbance to nesting, and the direct and indirect effects of pesticide use (Best, 1995; Rodgers and Wooley, 1983). Familiar wildlife species associated with agricultural land include the ring-necked pheasant, bobwhite quail, cottontail rabbit, meadowlark, white-tailed deer, killdeer, and barn owl. These species use cropland and grassland for nesting, feeding, and escaping predators, although it must be recognized that farming operations other than tillage take place on the acreage (Snider, Moore, and Subagja, 1985).

Conservation tillage use benefits wildlife mainly by leaving crop residue on the soil surface during spring and summer that may be used as cover (Brady, 1985). In a study conducted in the early 1980s, a greater abundance of invertebrates, birds, and mammals lived in conservation-tilled than in conventional-tilled corn fields in southern Illinois (Warburton and Klimstra, 1984). A greater diversity and density

TABLE 3.21. Benefits of Conservation Compliance Associated with Reduced Sheet and Rill Erosion

Crop Residue Cover	Number of Planted Acres (million acres)	Average Percent Residue Planned	Average Percent Residue Actual	Soil Loss Before[1] (t/a/yr)[4]	Soil Loss After[2] (t/a/yr)	Nominal Soil Erosion Benefits (million $) Range	Best[3]
0	22.87646	0.0	3.5	15.2	5.8	70.96276-206.4371	148.37670
< 15%	6.175330	9.4	31.6	9.9	5.0	9.985508-29.04875	20.87879
15-29%	18.01402	21.0	39.7	11.6	5.3	37.45115-108.9488	78.30696
30-39%	21.46073	31.1	45.8	14.7	5.5	65.15477-189.5411	136.23272
40-59%	17.18822	44.8	53.5	19.3	7.1	69.19977-201.3084	144.69041
> 59%	5.272614	71.5	73.2	20.3	6.4	24.1848-70.35776	50.56964
Total	90.98737	24.7	36.0	15.1	5.9	276.9394-805.6420	579.05520

Source: USDA, Natural Resources Conservation Service (1996, p. 43).

[1]Before the implementation of conservation compliance.
[2]After all conservation management systems and technical practices were implemented.
[3]Benefits deemed most likely.
[4]t/a/yr denotes tons per acre per year.

of birds nested in conservation-tilled fields than conventional-tilled fields in Iowa, and nest success was comparable to idle areas, such as fencerows and waterways (Baldassare et al., 1983; Basore, Best and Wooley, 1986; Castrale, 1985).[24] Additionally, increased residue cover tends to diversify rather than increase populations of small mammals (Young and Clark, 1983; Young, 1984). The relatively high residue associated with conservation tillage provides more, and ostensibly better, conditions for feeding. Tillage also influences the availability of insects (Martin, Zim, and Nelson, 1951). For example, it has been estimated that the amount of time a quail chick needs to satisfy its daily insect requirement is 4.2 hours on a no tillage soybean field, 11.1 hours on a no tillage cornfield, 22.2 hours on a conventional-tilled soybean field, and 25.1 hours on a conventional-tilled corn field (Conservation Technology Information Center, 1997). If weeds are controlled by herbicides, there is minimal disturbance to residue-nesting wildlife after planting. If weeds are controlled by cultivation as in the case of ridge tillage, there is a good chance of physical disturbance to residue-nesting wildlife.

Several factors influence how well conservation tillage will improve wildlife habitat, including the specific crop grown (Hoffman and Albers, 1984). Legumes (soybeans) are highly desirable because they are best able to support native grassland species (Ribaudo et al., 1990). Another factor is the quality of the cover established, in terms of height and density, which is related to the chosen mix of vegetation (Dunachie and Fletcher, 1970).

Wildlife recreationists are the primary beneficiaries from increases in wildlife populations associated with an increase in the use of conservation tillage. Hunters enjoy higher numbers and quality of species, while fishermen will encounter greater numbers of fish. Increased wildlife habitat also provides more opportunities for enjoying nonconsumptive activities such as bird watching, photography, and hiking.

Quantification of these sorts of benefits is difficult because there are no well-defined markets for the activities[25] and, hence, no available transaction prices. Some surveys, however, have endeavored to indicate the number of people engaged in various wildlife-associated recreational activities and the economic value they attach to them. These survey data can be used to infer something about the benefits of greater wildlife populations resulting from conservation tillage. One such survey is the *National Survey of Fishing, Hunting, and Wildlife-Associated Recreation—1991* conducted by the Fish and Wildlife Service (U.S. Department of the Interior, Fish and Wildlife Service, 1993). The survey consisted of a telephone interview and follow-up of 129,500 households nationwide of those who had fished, hunted, or engaged in a nonconsumptive wildlife-related activity in 1990. Among the values collected on the survey was the net economic value of a specific activity, where net economic value is defined as a participant's willingness to pay above what he or she actually spent to participate. The benefit to society is defined as the sum of the willingness to pay across all individuals. Estimates of the net economic value per person per year for trout fishing range from highs of $965, $584, and $485 in California, New York, and Arizona, respectively, to lows of $235 for Maine and $236 for Pennsylvania. The mean value across all states was $374 (Waddington, Boyle, and Cooper, 1994). For deer hunting, the net economic value per person per year ranges from $768, $744, $705, and $701 for Maryland, New Hampshire, Alabama, and Rhode Island, respectively, to lows of $168 for Iowa, $252 for Idaho, and $254 for Montana. The average over all the states was $490. Estimates of the net economic value per person per year for wildlife watching range from $763 in Indiana to $106 in North Dakota. Other states with high annual values include Alaska at $655 and Arkansas at $558. Iowa, West Virginia, and Maryland, with $136, $171, and $182, respectively, have relatively low net economic values per year. The average net economic value across all states was $278 (Waddington, Boyle, and Cooper, 1994).

The Reduction in Carbon Emissions
Associated with Conservation Tillage

An increase in conservation tillage has two identifiable effects with regard to carbon emissions. First, no tillage results in an increase in carbon retention in the soil because less organic matter is lost to oxidation from mixing of the soil, and soil temperatures tend to be lower, which slows oxidation from mixing of the soil. The lower soil temperature slows decomposition (Tate, 1987). Second, conservation tillage is more energy efficient than conventional tillage, requiring fewer machinery operations and, hence, less energy. Thus, carbon emissions are reduced because less fossil fuel (gasoline and diesel fuel) is used.

The amount and kind of crop residues have an effect on organic carbon levels in the soil (Rasmussen et al., 1980). Conservation tillage may increase the amount of organic carbon in the soil by providing an environment in which fungal decomposition is greater than bacterial decomposition. Fungal decomposition results in more recalcitrant decomposition products than bacterial decomposition (Holland and Coleman, 1987).

Historically, conventional tillage has resulted in losses of soil carbon (Hunt et al., 1996; Zobeck et al., 1995). Although this long-run effect of conventional tillage is clear, there are a number of unresolved questions with respect to modifying the processes that promote carbon loss from the soil. One is the relative importance of the different processes affected by tillage, such as soil disturbance, and the way tillage impacts changing carbon inputs and losses. An answer to this question is necessary before an accurate assessment can be made of the long-run effect of different tillage systems on the carbon level in the soil (Barker et al., 1996).

Although conventional tillage run does alter soil structure and increase the loss of soil carbon in the long run, the magnitude of these effects is a function of the intensity of tillage, the frequency of tillage, and the quantity and quality of fertilizer and organic residue returned to the soil (Rasmussen and Collins, 1991).

Conservation tillage has the potential for converting many soils from sources of atmospheric carbon to carbon sinks. The extent of the potential carbon sequestration, however, is highly variable (Donigian et al., 1994; Raich and Potter, 1997). This variability is a function of the crop grown, the cropping pattern, the type of soil, and climatic conditions. Given this variability, and given the fact that a substantial number of objective studies of carbon sequestration associated with conservation tillage for various crops,[26] cropping patterns, soil types, and climatic conditions is not yet available, it is difficult to assess the aggregate social benefit of reduced carbon emissions associated with increased use of conservation tillage.[27] Compounding the assessment problem is the fact that rates of carbon emissions from the soil are significantly correlated with temperature and precipitation and have a pronounced seasonal pattern. Hence, estimates of any aggregate effects must be based on studies incorporating climatic conditions and seasonal changes. The available data are simply not adequate to make a credible aggregate estimate, even though some have tried (e.g., Kern and Johnson, 1993).[28] More studies of the sort by Gilley and colleagues (1997) that assess the relative carbon content of different tillage practices on comparable soils are needed.

The second way that conservation tillage reduces carbon emissions is through reduced consumption of gasoline and diesel fuel (State of Illinois, 1997). Even with slightly more herbicide use in the first few years of some systems, conservation tillage is more energy efficient than conventional tillage (Frye, 1995). Quantifying the reduction in carbon emissions, however, due to reduced energy use is elusive. Farm equipment in U.S. agriculture consists of a mix of gasoline- and diesel-fuel-tractors (Uri and Day, 1992). A precise profile of this mix of gasoline- and diesel-fuel-using equipment, however, is not available. Since carbon emissions vary by type of fossil fuel and size of equipment, it is not possible to provide a quantitative estimate of any reduction in carbon emissions from a change in fossil fuel consumption associated with conservation tillage.

CONCLUSION

A review of relevant studies and results from the *Cropping Practices Survey* points to a lack of conclusive evidence that conservation tillage leads to higher yields (at least not in the short run). Moreover, the costs of inputs, including labor, fertilizer, pesticides, seeds, and machinery, are so dependent on site-specific factors that general inferences about production costs are not possible.

The evidence does lead to the following conclusions:

- If conservation tillage is used continuously, soil quality will improve, thus increasing the long-term productivity of the land.
- The use of conservation tillage does result in less of an adverse impact on the environment from agricultural production than does conventional tillage by reducing surface water runoff and wind erosion.

- Wildlife habitat will be enhanced to some extent with the adoption of conservation tillage.
- The benefits to be gained from carbon sequestration will depend on the soil remaining undisturbed.
- Further expansion of conservation tillage on highly erodible land will unquestionably result in an increase in social benefits, but the expected gains will be modest.

Several government policies can be used to encourage the adoption of conservation tillage if that is deemed desirable after weighing all the evidence. This is the subject of the following chapter.

APPENDIX

Technical Definitions

Universal Soil Loss Equation

The *Universal Soil Loss Equation* (USLE)[29] is

$$A = R * K * f(L,S) * C * P$$

where

A is the computed soil loss per unit area, expressed in the units selected for K and for the periods selected for R. In practice, these are usually selected so that they compute A in tons per acre per year;

R, the rainfall and runoff factor, is the number of the rainfall erodibility index units plus a factor for runoff from snow melt or applied water where such runoff is significant;

K, the soil erodibility factor, is the soil loss rate per erodibility index unit for a specified soil as measured on a unit plot, which is defined as a 72.6-foot length of uniform 9-percent slope continuously in clean-tilled fallow;

L, the slope length factor, is the ratio of soil loss from the field slope to that from a 72.6-foot length of uniform 9-percent slope continuously in clean-tilled fallow;

S, the slope steepness factor, is the ratio of soil loss from the field slope gradient to that from a 72.6-foot length of uniform 9-percent slope continuously in clean-tilled fallow;

C, the cover and management factor, is the ratio of soil loss from an area with specified cover and management to that from an identical area in tilled continuous cover; and

P, the support practice factor, is the ratio of soil loss with a supporting practice, such as contouring, strip cropping, or terracing, to that with straight-row farming up and down the slope.

Note that f(L,S) indicates a functional relationship between L and S.

Highly erodible land (HEL) is land determined to have an inherent erosion potential of over eight times its soil loss tolerance (T) level. Determination is made by calculating the erodibility index (EI) for both water and wind erosion. If the EI for either water or wind is greater than eight, then the soil is classified as HEL.

The *erodibility index* is a number showing how many times the inherent erosion potential is of the soil loss tolerance (T) level. For water (sheet and rill) erosion, the number is calculated as

$$EI = R * K * L * S / T$$

where

R, the rainfall and runoff factor, is the number of the rainfall erodibility index units plus a factor for runoff from snow melt or applied water where such runoff is significant;

K, the soil erodibility factor is the soil loss rate per erodibility index unit for a specified soil as measured on a unit plot, which is defined as a 72.6-foot length of uniform 9-percent slope continuously in clean-tilled fallow;

L, the slope length factor, is the ratio of soil loss from the field slope to that from a 72.6-foot length of uniform 9-percent slope continuously in clean-tilled fallow; and

S, the slope steepness factor, is the ratio of soil loss from the field slope gradient to that from a 72.6-foot length of uniform 9-percent slope continuously in clean-tilled fallow.

The *soil loss tolerance level* (T) is the maximum rate of annual soil erosion that may occur and still permit a high level of crop productivity to be obtained economically and indefinitely. Most values for cropland in the United States are between 3 and 5 tons per acre per year.

Wind Erosion Equation

The *Wind Erosion Equation* (WEQ)[30] is of the form

$$E = g(I, K, C, L, V)$$

where

E, the potential average annual soil in tons per acre per year, is the erosion that would occur from a field that is level, smooth, wide, bare, unsheltered, isolated, and having a climatic factor of 100 percent;

I is the soil erodibility factor;

K, the roughness factor, reflects the presence of ridges that, if at right angles to the wind, reduces wind erosion by reducing surface velocity and trapping particles;

C, the climatic factor, accounts for the influence of wind velocity and surface soil moisture;

L is the unsheltered travel distance along the prevailing wind erosion direction for the field or area to be evaluated; and

V is the vegetation cover.

The functional notation g(*) indicates that the relationship is nonlinear.

For wind erosion, EI is calculated as

$$EI = C * I / T$$

where

I is the soil erodibility factor; and

C, the climatic factor, accounts for the influence of wind velocity and surface soil moisture.

Chapter 4

Conservation Tillage:
The Role of Public Policy

- A number of policy tools are used to reduce soil erosion from agricultural lands in the United States, including education and technical assistance, financial assistance, land retirement, and conservation compliance requirements.
- Education and technical assistance by public and private sources can be effective in promoting the adoption of conservation tillage by farmers for whom that practice will be profitable.
- Financial incentives may be necessary to induce the voluntary adoption of conservation tillage by farmers for whom the practice would not be more profitable than conventional tillage but on whose land the use of conservation tillage would provide substantial off-site benefits.

Soil erosion associated with agricultural production practices can impose significant costs on both the public and private sectors. The adoption of conservation practices such as conservation tillage, contour farming, filter strips, grassed waterways, terracing, poly-acrylamide, and grasses and legumes in rotation can reduce soil erosion and the transport of sediments and chemicals to off-farm water bodies. Several public policies can be used to affect farmers' choices of production practices and technologies: education and technical assistance, financial assistance, research and development, land retirement, and regulation and taxes. Each policy has implications for agricultural profits and the allocation of public funds.

POLICIES DESIGNED TO AFFECT THE ADOPTION OF SPECIFIC PRODUCTION PRACTICES

Education and Technical Assistance

If a preferred practice would be profitable for a farmer but the farmer is unaware of its benefits, education efforts can lead to voluntary use of the practice. Educational activities generally take the form of demonstration projects and information campaigns in print and electronic media, newsletters, and meetings. Demonstration projects provide more direct and detailed information about farming practices and production systems and how these systems are advantageous to the producer (Bosch, Cook, and Fuglie, 1995). Information assumes an especially significant role in the case of new or emerging technologies (Saha, Love, and Schwart 1994). When adoption of a practice would lead to an increase in long-term profits, but either new skills are needed or farming operations must be adapted for the practice to produce the highest net benefits, technical assistance can be provided to those who choose to adopt. Technical assistance is the direct, one-on-one contact provided by an assisting agency or private company for the purpose of providing a farmer with the planning and knowledge necessary to implement a particular practice on the individual farm. Requirements for successful implementation vary among individual farms because of resource conditions, operation structure, and owner/operator managerial skill. Testing a practice on part of the farm enhances its potential for adoption (Office of Technology Assessment, 1990; Nowak and O'Keefe, 1995). Technical assistance is often critical, especially for practices that require a greater level of management than the farmer currently uses (Dobbs et al., 1995). Both education and technical assistance can be provided by either public or private sources, and both will induce adoption by farmers for whom the practice would be more profitable than the one they had been using.

Financial Assistance

Financial assistance can be offered to overcome either short- or long-term impediments or barriers to adoption. If the practice would be profitable once installed, but involves initial investment or transition and adjustment expenses, a single cost-share payment can be used to encourage the switch to the preferred practice. Transition and adjustment costs include lost production, increased risk, or increased management costs due to learning how to use the new production practice efficiently. Financial and organizational characteristics of the whole operation also may be a hindrance to adoption (Office of Technology Assessment, 1990; Nowak, 1992). When the practice would not be more profitable to the farmer than the current practice, but the environmental or other off-farm benefits are substantial, public funds could be allocated on an ongoing basis to defray the loss in profits to the farmer. Another form of financial incentive could be the granting of a tax credit for investment in a particular practice. From a public perspective, the optimal financial assistance rates are those which induce the adoption of desired practices at the least cost. Efficient rates would have to be set individually because farm and farmer characteristics vary widely (Caswell and Shoemaker, 1993). Therefore, for ease of implementation, most large financial assistance programs specify a uniform subsidy rate across resource conditions.

Uniform rates, however, invariably introduce production distortions. Because resource and production characteristics can vary widely, different farms may need different sets of practices to achieve the same environmental goal. A production system that is appropriate for one farm may be inappropriate for another. The effectiveness of a conservation system in controlling erosion depends on several factors, including the frequency, timing, and/or severity of wind and precipitation; the exposure of land forms to weather; the ability of exposed soil to withstand erosive forces; the plant material available to shelter soils; and the propensity of production practices to reduce or extenuate erosive forces. An efficient financial assistance program would have a list of eligible management practices that included all alternatives appropriate for each farm. Cost-share and incentive payment policies are based on the fact that targeted farmers would not voluntarily adopt the preferred practice, but the public interest calls for the practices to be used more widely. Financial assistance is not a substitute for education and technical assistance. Even with financial assistance, a farmer will not adopt a technology if he or she is unfamiliar with it.

Research and Development (R&D)

Research and development policies can be used to enhance the benefits of a given production practice. The objective of the research would be either to improve the performance or to reduce the costs of the practice. Data gathering and analysis, as well as monitoring, also contribute to R&D by providing information necessary to assess the determinants of adoption and the effectiveness of practices in achieving public goals. In addition, R&D funds could be allocated to ensure that the practice is adaptable for more circumstances. R&D is a long-term policy strategy with an uncertain probability of success, but it may also reap the greatest gains in encouraging the voluntary adoption of a preferred technology because it can increase the profitability of the practice for a wider range of potential adopters.

Land Retirement

The policy that has the largest impact on farmers' choice of practices or technologies is land retirement. The underlying premise is that large public benefits can be gained by radically changing agricultural practices on particular parcels of land and that changes in individual practices would not provide sufficient social benefits. For an individual to voluntarily agree to allot the land for conservation uses, he or she would expect compensation in an amount at least as great as the lost profits from production. The payment mechanisms that can be used to implement land retirement strategies are lump sum payments or annual "rental fees." The former are often referred to as easements, whereby the farmer's right to engage in nonconservation uses is purchased by the public sector for a specific period. Payment to an individual to retire land would result in a voluntary change in practices.

Regulation, Taxes, and Tax Incentives

If voluntary measures prove insufficient to produce the changes in practices necessary to achieve public goals, regulation is a policy that can be used. The use of certain practices could be prohibited, taxed, or made a basis for withholding other benefits. Preferred practices could be required or tax incentives offered to promote their use, thereby offsetting some of the cost of new conservation tillage equipment. Point sources of pollution have been subject to command-and-control policies for many years. Recognized inefficiencies are associated with technology-based regulations because the least-cost technology combination to meet an environmental goal for an individual may not be permitted.[1] It has been assumed that such loss in efficiency is made up for by ease of implementation.

Policies to control soil erosion on agricultural lands have been administered mainly by the U.S. Department of Agriculture. The following section describes the policies used by the USDA to promote soil conservation.

U.S. DEPARTMENT OF AGRICULTURE SOIL CONSERVATION POLICIES

The conservation and related water quality programs administered by the U.S. Department of Agriculture primarily have been designed to induce the voluntary adoption of conservation practices. The USDA has used a number of policy tools, including on-farm technical assistance and extension education; cost-sharing assistance for installing preferred practices; rental and easement payments to take land out of production and place it in conservation uses; R&D for evaluating and improving conservation practices and programs; and compliance provisions that require the implementation of specified conservation practices or the avoidance of certain land use changes if a farmer wants to be eligible for federal agricultural program payments. Regulatory and tax policies have not been part of the traditional voluntary approach of U.S. conservation programs. The USDA policy has been to decrease government involvement in farm operations (Reichelderfer, 1990).

The USDA conservation programs are closely tied to state and local programs. Federal and state agencies cooperate with a system of special purpose local (county) conservation districts that are authorized by state law to provide education and technical assistance to farmers and with county Agricultural Stabilization and Conservation (ASC) committees to handle cost sharing (Libby, 1982).[2] The system ensures that financial support and technical assistance are focused on the set of problems relevant to the geographic region and the national interest. The adoption of an alternative production practice generally does not occur as a consequence of any one specific assistance program (Missouri Management Systems Evaluation Area [MSEA], 1995). The USDA has a memorandum of understanding with each conservation district to assist in carrying out a long-term program.[3] Conservation districts have proven to be practical organizations through which local farmers and the federal government can join forces to carry out needed soil conservation practices (Rasmussen, 1982). The demand for information has changed over time. Not long ago, an extension agent was the primary source of information on new technologies (van Es, 1984; Hefferman, 1984). Now, however, farmers are relying on many additional sources of information, including newspapers, magazines, agrichemical dealers, crop consultants, and the Internet.

The U.S. Department of Agriculture, Natural Resources Conservation Service, formerly the Soil Conservation Service, provides technical assistance to farmers and other land users, including local, state, and federal agencies that manage publicly owned land. NRCS helps district supervisors and others to draw up and implement conservation plans.

Providing federal cost-sharing assistance to farmers for voluntary installation of approved conservation practices is the responsibility of state and county ASC committees. Through the Agricultural Conservation Program, funds were allocated among the states through the state ASC committees on the basis of soil and water conservation needs. ACP practices eligible for cost sharing were established by a national review group representing all USDA agencies with conservation program responsibilities, the U.S. Environmental Protection Agency, and the Office of Management and Budget. The practices were designed to help prevent soil erosion and water pollution from animal wastes or other nonpoint sources, to protect the productive capacity of farmland, to conserve water, to preserve and develop wildlife habitat, and to conserve energy (Holmes, 1987).[4]

The Secretary of Agriculture can also target critical resource problem areas for financial and technical assistance based on the severity of the problem and the likelihood of achieving improvement. Highly erodible and/or environmentally sensitive cropland has recently been targeted because the greatest net social benefits were expected to be associated with a reduction in soil erosion on these land classifications. Targeting, however, will not guarantee that the net benefits (public and private) of any conservation practice will be positive because net benefits are a function of site-specific factors.

The policy to take land out of production and place it into conservation uses was first used in the Soil Bank Program of the 1950s and has been significantly increased in the current Conservation Reserve Program (CRP). The Conservation Reserve Program provides for the USDA to enter ten- to fifteen-year agreements with owners and operators to remove highly erodible and other environmentally sensitive cropland from production. Along with conservation, the CRP originally had a secondary objective of reducing surplus crop production (Osborn, 1996). The more recent emphasis in CRP, however, has been to provide environmental benefits rather than to control the supply of commodities.

Most of the highly erodible land (HEL) contracted into the CRP had suffered much erosion, organic matter loss, and structural deterioration while it was in cultivated crop production.[5] When lands are returned to grass, their structure and organic matter improve and tend to approach the structure and organic matter content of the original grassland soils (Gebhart et al., 1994). The degree of soil improvement from ten years of grass is a function of site-specific factors. As a general rule, the greater the amount of soil structure deterioration from past cultural practices, the more likely that grass management will improve the soil's characteristics. Rasiah and Kay (1994) found that if soils had higher levels of organic matter and other stabilizing materials at the time of grass introduction, the time required for soil structure regeneration was reduced. Soils in the CRP typically fit into the category of degraded soils whose organic matter is lower than that of surrounding soils because they were primarily allowed into the program based on their highly erodible classification (Lindstrom et al., 1992; Barker et al., 1996).

CRP contracts are beginning to expire, so farmers have the option to return the land to crop production. For land that would be returned to production, the improvements in soil quality and erosion reduction gained during the CRP contracts will be rapidly lost if conventional tillage is used. A concern is whether it will be possible to maintain the benefits derived from ten years of grass. Several options exist for post-CRP land:

1. Keep HEL land in the CRP. This would allow the soil in the program to continue to improve and maintain the erosion benefits.
2. Subsidize a rotation that involves four years in grass production followed by four years in grain production.
3. Lower CRP payments to keep the land in grass, but allow grazing or haying on the land. This proposal has met with considerable opposition by farmers who have land already in hay production and who object to subsidized hay production that would compete with their commodity. (Schumacher et al., 1995)

The focus of current conservation research has been on the development of environmentally acceptable and sustainable production practices. Goals of this research are to gain a better understanding of how different soils respond to tillage, what amount of tillage is necessary for optimum crop growth, and what combination of mechanical, chemical, and biological practices are needed to create environmentally sustainable crop production. Many conservation practices have been evaluated by the land grant university experiment stations (Moldenhauer, Kemper, and Langdale, 1994). USDA has funded several major surveys to provide data to assess the extent and determinants of adoption for particular production practices across a wide range of crops and regions.[6] A more in-depth discussion of conservation-related R&D can be found in Karlen (1990).

The major shift in U.S. conservation policy came in the Food Security Act of 1985 in the form of conservation compliance (Heimlich, 1991). This provision provided farmers with a basic economic incentive to adopt conservation tillage practices or another acceptable plan because agricultural program payments were linked to the adoption of an acceptable conservation system on highly erodible land. Although meeting the conservation provisions remains voluntary, a farmer who wants to receive certain agricultural program payments and whose cropland is designated as HEL has no choice but to implement an acceptable conservation plan (Crosswhite and Sandretto, 1991). The conservation compliance provision was innovative because it linked farm program payments (private benefits) to conservation performance (social benefits).[7] In 1982, cultivated HEL accounted for almost 60 percent of the total erosion on U.S. cropland in terms of tons per acre per year, although it accounted for only 40 percent of total planted acreage (Magleby et al., 1995). Requirements for conservation compliance were applied to HEL previously cultivated in any year between 1981 and 1985. Conservation compliance required farmers producing crops on HEL to implement and maintain an approved soil conservation system by 1995.

Acceptable conservation plans specify economically viable conservation systems designed to reduce soil erosion. Conservation systems are composed of one or more conservation practices. The 1995 status review of approved conservation systems by the Natural Resources Conservation Service provides the first assessment of fully implemented conservation systems under conservation compliance (USDA, Natural Resources Conservation Service, 1996). Although the 1995 status review found over 4,000 different conservation systems (combinations or practices) applied throughout the United States, four conservation systems involving conservation cropping sequences, crop residue use, or a combination of these practices with conservation tillage accounted for approximately half of planted HEL acreage (see Table 4.1).

Regional differences in the adoption of specific conservation systems can be illustrated by a comparison of the conservation plans from Iowa, North Carolina, North Dakota, and Oklahoma. In the relatively homogeneous Northern Plains, there are few economically viable alternatives to a wheat/fallow rotation. Consequently, in North Dakota, a conservation crop sequence/crop residue management system was part of nearly all conservation systems used on planted HEL in 1995 (see Table 4.2). Analogously, in the Southern Plains, wheat is the predominant crop, with few economically viable alternatives. In Oklahoma, most approved conservation systems consist of a single technical

TABLE 4.1. Conservation Management Systems and Technical Practices Being Applied on Cultivated HEL Subject to Compliance (Excluding CRP)— 1995

Item	Acreage	Percent of Cultivated HEL[1]
Management Systems		
Conservation cropping sequence/crop residue use	27,443,973	30.2
Conservation cropping sequence/conservation tillage	9,081,148	10.0
Conservation cropping sequence only	6,249,209	6.9
Crop residue use only	4,041,388	4.4
Conservation cropping sequence/conservation tillage/ grassed waterways	2,027,771	2.2
Conservation cropping sequence/conservation tillage/ grassed waterways/terrace	1,958,476	2.2
Conservation cropping sequence/contour farming/ crop residue use/terrace	1,896,080	2.1
Conservation cropping sequence/crop residue use/ wind strip cropping	1,768,605	1.9
Conservation cropping sequence/contour farming/ crop residue use/grassed waterways/terrace	1,665,697	1.8
Conservation cropping sequence/conservation tillage/ crop residue use	1,602,604	1.8
Total, 10 most frequently used systems	57,734,951	63.5

Technical Practices[2]

Technical Practices[2]		
Conservation cropping sequence	75,632,767	83.1
Crop residue use[3]	48,294,496	53.1
Conservation tillage[3]	28,477,584	31.3
Contour farming	18,046,999	19.8
Terrace	12,868,684	14.1
Grassed waterway	10,842,932	11.9
Field border	4,442,198	4.9
Wind strip cropping	3,508,340	3.9
Cover and green manure	3,169,983	3.5
Surface roughing	3,018,871	3.3
Grasses and legumes in rotation	2,424,281	2.7
Strip cropping-contour	1,699,477	1.9
Critical area planting	1,545,287	1.7
Pasture and hay land management	1,126,426	1.2

Source: USDA, Economic Research Service (1997, p. 127).

[1]Based on 91 million acres of cultivated HEL subject to compliance.
[2]Because many conservation systems include multiple technical practices, percentages will sum to more than 100.
[3]Conservation tillage and crop residue management are frequently combined and reported as a single practice, conservation tillage.

TABLE 4.2. Technical Practices Included in Conservation Plans in Iowa, North Carolina, North Dakota, and Oklahoma—1995

Technical Practice	Iowa	North Carolina	North Dakota	Oklahoma
	Percent of conservation plans[1]			
Conservation crop rotation	87.1	82.0	99.0	9.9
Conservation tillage	79.2	30.6	0.4	3.5
Residue management	0.7	50.5	98.4	92.3
Contour farming	44.4	24.3	*	5.4
Strip cropping – field border	32.3	15.0	*	*
Strip cropping – contour	2.3	*	*	*
Strip cropping – field	*	5.0	*	*
Strip cropping – wind	*	*	0.6	0.3
Grassed waterway – retired[2]	24.9	21.9	0.7	8.2
Grasses and legumes in rotation	1.0	7.2	*	*
Cover and green manure crop	*	5.1	1.5	0.3
Conservation cover – retired[2]	*	13.6	3.0	0.5
Critical area planting – retired[2]	0.8	4.3	0.1	0.6
Terrace	13.4	1.2	*	0.2
Pasture and hay land management	13.7	5.9	0.2	22.5
Pasture and hay land planting	1.3	6.3	0.4	0.3

Source: USDA, Economic Research Service (1997, p. 129).

*Indicates less than 0.1 percent.
[1]Because many conservation systems include multiple practices, percentages will sum to more than 100.
[2]Retired indicates land taken out of production.

practice: crop residue management. Both the number of viable conservation systems and the number of systems required to control erosion effectively are greater in areas with relatively greater climatic and geographic variability. Iowa produces predominantly corn and soybeans and has a higher average rainfall and more varied topography than North Dakota and Oklahoma. As a result, a larger number of conservation systems are used in Iowa, most frequently conservation cropping sequences and conservation tillage. North Carolina has a variable topography with diverse soils and precipitation patterns and produces a large number of different agricultural commodities, including wheat, corn, soybeans, cotton, sorghum, and tobacco. Consequently, the conservation systems used in North Carolina are even more varied than they are in Iowa.

The most recent manifestation of agricultural program policy is the Federal Agriculture Improvement and Reform Act of 1996. It modifies the conservation compliance provisions of the Food Security Act of 1985 to provide farmers with greater flexibility in developing and implementing conservation plans, in self-certifying compliance, and in obtaining variances for problems affecting application of conservation plans. Producers who violate conservation plans or fail to use a conservation system on highly erodible land, risk loss of eligibility for many payments, including production flexibility contract payments. An important aspect of this act is that in self-certifying compliance, there is no requirement that a status review be conducted for producers who self-certify (Nelson and Schertz, 1996). The FAIR Act also does not differentiate between previously cultivated and uncultivated land, thereby eliminating the sodbuster program.[8] Newly cropped highly erodible land may use conservation systems other than the systems previously required under the sodbuster program.

Additionally, the FAIR Act established a new program, the Environmental Quality Incentive Program (EQIP), that incorporated the functions of ACP and some other environmental programs, designed to encourage farmers to adopt production practices that reduce environmental and resource problems. The acceptable plans will improve soil, water, and related natural resources, including grazing lands, wetlands, and wildlife habitat. EQIP must be carried out to maximize environmental benefits provided per dollar expended. During 1996 to 2002, the Secretary of Agriculture will provide technical assistance, education, and cost sharing to producers who enter into five- to ten-year contracts specifying EQIP conservation programs. Based on the historical experience of the ACP, very little of the EQIP funds are likely to be targeted at farmers using conservation tillage.[9]

POLICY IMPACTS ON THE ADOPTION
OF CONSERVATION TILLAGE

The adoption of any agricultural technology is a function of many things. As described earlier, the choice of practice is influenced by farm and farmer characteristics, attributes of the technology, economic conditions, and public policies. For the adoption of conservation tillage, some factors are easier to influence than others. For example, ownership characteristics can also influence the adoption of conservation tillage. Owner-operators are more likely to have greater flexibility to adopt conservation tillage than non-owner-operators who must often get approval from the owner before making production practice changes. Conservation tillage may tax the managerial skill of the operator (Nowak, 1992). Farmers typically make production decisions within a short time frame, a fact that may discourage investment in measures that increase returns only over the long run, as may be the case with conservation tillage (Office of Technology Assessment, 1990; Tweeten, 1995). Risk-averse producers do not look favorably upon practices that are perceived as being too risky, and in many situations, conservation tillage is more risky relative to conventional tillage because of the timing and managerial aspects and greater variability in yields. Also, access to capital may depend on risk. A relatively greater chance exists that something might go wrong with conservation tillage, and then net returns will fall (Fox et al., 1991).

Most conservation policies attempt to influence the use of conservation tillage through demonstrating or ensuring that net benefits of adoption are positive. The policies that have had the greatest impact to date have been those focusing on education and technical assistance. The management complexities associated with conservation tillage relative to conventional tillage are considerable.

Farming using conservation tillage requires a different approach to soil preparation, fertilizer application, and weed and insect control. Moreover, conservation tillage systems must be designed according to the unique conditions of the region and the specific needs of the individual farmer. Consequently, it is not feasible to design a conservation tillage system that can be applied at all locations across the United States, or even within a single region. A successful conservation tillage system must be developed from a whole-system perspective. Simply stopping tillage with no other changes in the cropping system increases the potential for problems and failure (Schumacher et al., 1995).

Reasons posited for the inability of farmers to adopt conservation tillage even under favorable economic conditions with positive net private benefits include a lack of information, a high opportunity cost associated with obtaining information, complexity of the production system, a short planning horizon, inadequate management skills, and a limited, inaccessible, or unavailable support system (Nowack, 1992). Westra and Olson (1997) conducted a survey of farmers in Scott and Le Sueur Counties in Minnesota. They found that several noneconomic factors impact the decision to use conservation tillage, including whether the farmer perceives he or she has the requisite management skill for conservation tillage use and whether conservation tillage fits in with the farmer's production goals and physical setting of the farm. One surprising result from the survey was that 47 percent of the respondents indicated that they knew nothing about conservation tillage. The surveyed counties, however, have little acreage designated as HEL.

Education and technical assistance programs funded by the government can be effective in increasing the use of conservation tillage (Logan, 1990). Dickey and Shelton (1987), for example, discuss the education programs in eastern Nebraska that targeted specific areas that were susceptible to relatively high soil erosion rates. The program succeeded in increasing the use of conservation tillage by 20 percent and in reducing soil erosion by 20 percent in the target areas.

Education and technical assistance to mitigate the impediments to use of conservation tillage associated with real or perceived management inadequacies can come in a variety of forms, not just through extension education and county agents who work for the government. A large amount of information is available through land grant universities, agrichemical dealers, and independent crop consultants. For example, the Washington State University (WSU) Agriculture Extension Service has prepared the *Pacific Northwest Conservation Tillage Handbook,* which addresses virtually all management issues associated with the adoption of conservation tillage in the Pacific Northwest (WSU Agriculture Extension Service, 1997). The Agricultural and Biosystems Engineering Department at Iowa State University has produced the publication *Conservation Tillage Systems and Management,* which addresses management problems associated with conservation tillage in the Midwest (Midwest Plan Service, 1992). Private groups also provide management information. The Conservation Technology Information Center prepares publications such as *Conservation Tillage: A Checklist for U.S. Farmers* (Conservation Technology Information Center, 1997), which addresses the weed, insect, disease, and nutrient management issues associated with conservation tillage. Thus, quite a few publications are available in hard copy or through the Internet that address management issues associated with conservation tillage.

Agricultural input supply dealers and crop consultants are also good sources of information on the management complexities associated with conservation tillage and how to deal with them. Fertilizer and pesticide dealers and crop consultants have consistent access to farmers and, consequently, have the potential for exercising great influence on the tillage system used (Center for Agricultural Business, 1995). Until recently, there was little information on the influence of dealers and consultants in tillage decisions. Recent survey instruments include questions about whether decisions are influenced by these sources, but still no objective studies quantify the impact of dealers and crop consultants in assisting farmers on the use of conservation tillage with various other management problems (Wolf, 1995). Education and technical assistance efforts will succeed in inducing adoption of conservation tillage only by those farmers who can be shown that they will reap net benefits in the long run. For farmers with production or resource characteristics for which conservation tillage is not profitable, education and technical assistance will not be a sufficient inducement to adopt.

Another impetus to increasing the use of conservation tillage would be the return of Conservation Reserve Program acreage to crop production. The adoption of conservation tillage appears to provide the greatest potential for achieving positive net private benefits, while retaining most of the soil quality improvement achieved during the CRP and coincidentally mitigating soil erosion relative to conventional tillage (Veseth et al., 1997). The precise strategy to follow, however, depends on weed problems, water availability, seasonal workload, the economics of crop options, and so forth, in other words, site-specific factors.

Conservation tillage has proven to be an important component of conservation plans developed under conservation compliance provisions (see Tables 4.1 and 4.2). Adherence to a conservation plan often entails some costs for most farmers. Often, the plan requires the purchase of new equipment to implement a conservation tillage system and human capital expenses to learn new production practices. Farmers who consider adopting conservation tillage in the future must consider these sorts of costs in making their production decisions. Costs and benefits of the adoption of conservation tillage to society as a whole are much more difficult to measure.[10] Although higher production costs and reduced output would ostensibly lead to higher consumer costs for food and fiber, there is scant evidence that this, in fact, has occurred. If it has, it is relatively modest (Young, Wallace, and Kanjo, 1991).

A farmer whose cropland is highly erodible and who adopts conservation tillage as part of a conservation plan will benefit by controlling the rate of soil erosion, thereby maintaining the long-term productivity of the soil. The significance of this benefit depends on a number of elements, including current topsoil depth, erosion rates, and the rate at which the farmer discounts benefits in future years (Tracy, 1993). Conservation tillage has the added benefit, unlike many other conservation practices, that it can lead to an increase in soil organic matter, an increase in soil moisture, reduced soil compaction, and so forth. All of these have the potential of enhancing soil quality, leading to relatively greater yields in the long run and an increase in the value of a farmer's primary asset, the land.

CONCLUSION

Soil conservation programs in the United States have traditionally employed four major tools to encourage the adoption of preferred practices and technologies: education and technical assistance, financial assistance, crop acreage diversion programs, and, more recently, conservation compliance.

Conservation tillage is one of the practices that can be used to reduce soil erosion and maintain productivity of agricultural land. Education and technical assistance are effective mechanisms for increasing the use of conservation tillage by farmers for whom the technology will be profitable. Given the management complexities associated with the use of conservation tillage, the availability of education and technical assistance is critical. Education and technical assistance, however, do not have to come exclusively from the public sector. The private sector, through commodity groups, input suppliers, and crop consultants, can have a strong influence on decisions made by farmers.

For farmers who would not gain from the adoption of conservation tillage, financial incentives would be necessary to induce the voluntary change of practices. Such a policy is in the public interest if the off-site benefits of adoption outweigh the costs of the financial assistance program.

Summary

Soil erosion has both on-farm and off-farm impacts. The reduction of soil depth reduces the productive life of the agricultural land, and the transport of sediments and adhering chemicals can degrade streams, lakes, and estuaries. Many individual conservation practices and combinations of practices can be used to reduce soil erosion. The system used will depend on farm and farmer characteristics, attributes of the practices, economic conditions, government policies, and the value of the environmental resources affected by erosion. To be functional, the system must be manageable and economically viable (USDA, Natural Resources Conservation Service, 1997b). Conservation tillage is only one of the conservation practices that may be considered. On highly erodible land (HEL), as well as non-highly erodible land (NONHEL), the adoption of conservation tillage can offer substantial benefits to farmers by sustaining productivity and to the public by reducing sediment and chemical loadings in valued water bodies. Farmers also gain from improved soil qualities, such as increased organic matter and water-holding capacity, and the public will benefit from improvements in wildlife habitat and carbon sequestration.

Soil conservation policies have existed in the United States for more than sixty years. Initially, these policies focused on the on-farm benefits of keeping soil on the land and increasing net farm income. Beginning in the 1980s, however, policy goals increasingly included reductions in off-site impacts of erosion. Conservation tillage was one of the practices that was included in the suite of best management practices recommended within conservation programs. The Food Security Act of 1985 was the first major legislation explicitly to tie eligibility to receive agricultural program payments to conservation performance. The adoption of conservation tillage on highly erodible land increased significantly as the conservation compliance provisions of the 1985 Food Security Act took effect. The Federal Agriculture Improvement and Reform Act of 1996 modifies the conservation compliance provisions by providing farmers with greater flexibility in developing and implementing conservation plans. Noncompliance for those on HEL can result in loss of eligibility for many payments, including production flexibility contract payments.

The current definition of conservation tillage was not developed until 1984. Therefore, it is difficult to determine long-term trends in use and the impact of specific factors on these trends. The emphasis now is on leaving crop residue on the soil surface after planting. Using a broad definition, it was shown that the use of conservation tillage increased from 2 percent of planted acreage in 1968 to nearly 36 percent of planted acreage in 1996.

Use of conservation tillage varies significantly by crop and by geographic region. The practice is used mostly on corn, soybeans, and small grains. Of the major field crops, cotton has the lowest proportion of acreage under conservation tillage. Several important crops (peanuts, potatoes, sugar beets, tobacco, and vegetables) require production practices that preclude the use of conservation tillage. Therefore, policies to encourage adoption on acreage growing these crops will be less successful. Regional differences in the adoption rates for conservation tillage are also substantial. For example, Kentucky has 73 percent of its cropland acreage in conservation tillage versus less than 10 percent for most of New England.

The private benefits to farmers from use of conservation tillage depend on many site-specific factors. A survey of yield and cost differences between conservation tillage and conventional tillage systems shows inconclusive results. In some circumstances, one technology is more profitable, and in other situations, another system would be the economic choice for the farmer. The larger the difference a farmer perceives in net benefits between conservation and conventional tillage systems, the higher the probability that he or she will choose the more profitable option. The effects of adoption on some input uses also is not clear. Although herbicide use in conservation tillage seems to increase in the short run, agronomists hypothesize that herbicide usage will decrease in the long run. Recent surveys have shown that most farmers have been using conservation tillage for a relatively short time, however, so there is little experience from which to test the hypothesis.

The public benefits to be gained by use of conservation tillage also depend on site-specific factors. The gains from reducing sediment and chemical transport from highly erodible land can be significant, but off-site benefits from conservation tillage usage on non-highly erodible land may be relatively small.

Use of conservation tillage on an additional 48 million acres of HEL and 166 million acres of non-highly erodible cropland would yield a total reduction in erosion of 322 million tons per year.[1] Additionally, the benefits of converting the remaining HEL on which no conservation system is currently used to conservation tillage are relatively modest—approximately $50 million. Improvements in wildlife habitat and possible carbon sequestration would yield additional, but unquantifiable, benefits. The sum of all these benefits can be changed substantially by the total complement of production practices, not just those directly related to tillage activities.

Public education and technical assistance policies to promote the use of conservation tillage have been very effective. University and private sources of assistance also influence farmers' technology choices when the conservation tillage practices are profitable. When the adoption of conservation will provide significant public benefits, but would not be profitable for the farmer, financial incentives may be necessary to elicit a voluntary change in farmer practices.

Notes

Chapter 1

1. Sheet and rill erosion is the most common form of agricultural soil erosion. It occurs when raindrops or irrigation detach soil particles from the soil surface and transport them in thin sheets of water moving across unprotected slopes. As runoff water becomes concentrated, first into rills and then into separate channels, it begins to cut gullies, removing larger volumes of soil.

Chapter 2

1. The definition was developed through a joint effort involving the Soil Conservation Service, the Economic Research Service, and the Agricultural Research Service of the U.S. Department of Agriculture, the Conservation Technology Information Center, and private sector participants.

2. The CTIC continues to use this definition for the sake of long-term trend analysis. The Natural Resources Conservation Service (the former Soil Conservation Service) no longer uses this definition.

3. It was later renamed the Conservation Technology Information Center.

4. In 1996, the CPS and the Farm Costs and Returns Survey (FCRS) were combined into the Agricultural Resource Management Study (ARMS) survey. The intention is to provide information on production practices and costs and returns. The ARMS survey provides more information on the costs of production and production characteristics associated with different tillage practices than did the CPS, thereby providing the ability to identify the factors affecting the conservation tillage adoption decision.

5. Schertz estimates conservation tillage for the period 1968 through 1981. His data are taken from a variety of sources, and adjustments are made for changes in the definition of conservation tillage.

6. Nearly 47 percent of Indiana's planted acreage is in corn. Also, more than 7 percent of total corn planted acreage in the United States is in Indiana.

7. This examination of the trend in the use of conservation tillage is not meant to be exhaustive. Rather, it illustrates general trends in conservation tillage in U.S. agriculture. The reader interested in the many specific details for individual states and/or categories of conservation and conventional tillage practices is referred to Conservation Technology Information Center's *National Crop Residue Management Survey*, Bull and Sandretto (1996), and USDA, Economic Research

Service (1997) for more details on specific geographic aspects of the use of conservation tillage.

8. Consistent data by land class designation are currently available only for this period.

9. Cotton data are not reported. In 1995, conservation tillage was used on less than 2 percent of highly erodible cotton acreage and on less than 1 percent of non-highly erodible acreage.

10. This factor will be dealt with in greater detail in Chapter 3.

11. Because of the nature of the data available and their limitations, however, it is generally not possible to quantify precisely the impact that these factors have on conservation tillage use. Among the more serious data limitations are the availability of only a relatively short time series, incomplete measurement of all of the relevant variables for the times series that are available, measurement error in the data, absence of panel data, and difficulty in separating the impact of government policy (conservation compliance) from varying climatic and economic conditions. With regard to the first issue, consistent data on conservation tillage date only to the late 1980s. Six or seven years' worth of annual observation is typically inadequate to estimate a formal structural model when more than just a few factors affect the dependent variable (the relative use of conservation tillage). Of the time series data that are available, information on coincident climatic conditions and relevant economic variables are not collected in any sort of usable way. The changing definition of conservation tillage over the longer period has introduced measurement error of the dependent variable. The absence of panel data makes it impossible to monitor the behavior of any group of farmers over time. Consequently, because of these data limitations, in the review that follows, many of these factors affecting conservation tillage use will be discussed, but it is not feasible to quantify their relative importance in influencing the trend in the use of conservation tillage.

12. The last year for which the FCRS collected data sufficient to permit estimation of a production function was 1987. In subsequent years, the survey was scaled back due to concern over respondent burden.

13. Slope is one component of the universal soil loss equation and is used in computing the erodibility index. The value of the erodibility index determines whether the cropland is classified as HEL. See the appendix to Chapter 3 for details.

14. The soil types are not identified.

15. For a more extensive review, see Fox and colleagues (1991) and Roberts and Swinton (1996).

Chapter 3

1. Note that the issue of equity, an important part of the historical justification for soil conservation programs (Conway, 1994; Strohbehn, 1986), is not addressed in this report. Although soil conservation programs originated during the 1930s and had employment and income support objectives (Rasmussen, 1982),

comparing conservation tillage to conventional tillage with respect to equity goals involves too many subjective factors to be included here.

2. On only 12 percent of the planted acreage for major field crops in the United States in 1995 was a moldboard plow used under conventional tillage (USDA, Economic Research Service, 1997).

3. An integral part of the private conservation tillage adoption choice is the decision at what level to control soil erosion. A formal statement of this decision has been developed in a number of places (Foltz, Martin, and Preckel, 1995; McConnell, 1983; Miranowski, 1993). The essence of the statement is that in selecting the most economically efficient production practice where soil erosion control is an argument in the objective function, efficiency is achieved when soil loss occurs at the level where the value of the returns associated with additional soil loss equals the implicit cost of losing the additional soil (i.e., marginal forgone future returns). Thus, higher levels of yield loss per unit of soil loss will lower total soil loss during a given period. If a farmer places a relatively higher value (i.e., greater forgone yield) on a unit of topsoil, a greater effort will be made to reduce soil loss than with a lower value. Also, if agricultural output prices are increasing or are expected to increase over time, a more soil-conserving practice will be adopted, all other things being equal, since the value of returns associated with soil loss will be higher.

4. The revenue a farmer expects to receive from producing a crop is a function of expected yield and expected output price. The latter component is not affected by the choice of tillage practice and so it is not discussed here. The higher the output price, however, the more influential yield differences between tillage practices become.

5. The yield benefits of the various types of conservation tillage vary. No tillage will typically improve soil conditions much more rapidly than does mulch tillage or ridge tillage. Thus, the research comparing conservation tillage to conventional tillage has focused primarily on conventional tillage relative to no tillage (Zero Tillage Farmers Association, 1997).

6. The rills and gullies on the surface and sand and soil deposits on the bottomlands resulting from the heavy rains in 1993 forced farmers to till the soil more. Consequently, most of the benefits in terms of improved soil characteristics associated with conservation tillage were lost. The costs in the form of reduced net farm income for farmers who adopted conservation tillage were ex post facto greater than they had anticipated because the expected yield gain (if any was anticipated) from the use of conservation tillage could not be realized in the immediate future.

7. The exception is spring wheat yields in 1992. Note that here, and subsequently, tests of statistical significance are conducted at the 5 percent level. In drawing conclusions about statistical significance of the difference in the means of two samples, information on the sample size is needed. In the case of the *Cropping Practices Surveys*, the sample sizes are large enough so that the asymptotic limits of the relevant distributions can be used. Consequently, sample size information is not reported repeatedly. By way of example, however, the sample size

for the full corn sample in 1995 was 2,745. For soybeans, it was 2,085, and for winter wheat, it was 1,936.

8. Although the CPS data are not adequate to enable estimation of complete production functions, since data on many of the relevant factors impacting production were not collected, it is still possible to perform an analysis of variance to determine if, at an aggregate level, an identifiable relationship exists between yield and the tillage practice used. Although an analysis of variance will permit a determination of the presence of a relationship, it will not allow a measurement of yield differentials associated with different tillage practices and attributable to these practices.

9. The standard error of the mean is in parentheses.

10. As before, the standard error of the mean is in parentheses.

11. The standard error is in parentheses.

12. A companion piece repeats the conclusions of this study. See Ciba-Geigy Corporation (1992).

13. The conditions under which the field experiments are performed, and from which averages are computed, are critical. Basta, Huhnke, and Stiegler (1997) explore this issue.

14. A more recent development in terms of measuring soil loss potential is the work of the Water Erosion Prediction Project (WEPP). Because it is still under development and not completely operational, it will not be used here. The interested reader is referred to Becker, 1997.

15. An analysis of the structure of these relationships can be found in Wittmuss (1987) and Uri and Hyberg (1990).

16. The latest version of USLE is the revised universal soil loss equation (RUSLE). It is not used here because it was not used in developing the data for the 1992 Natural Resources Inventory. RUSLE is an improvement over USLE in that it incorporates more data (from different crops and cropping patterns, different locations, for forest and rangeland erosion), it corrects errors in the USLE analysis and fills gaps in the original data, and it possesses increased flexibility that allows modeling of a greater variety of systems and alternatives (Yoder and Lown, 1995; Renard et al., 1996). For the analysis presented here, the results would be substantially the same using either measure.

17. Annual updates to the NRI were prepared for selected subsamples in 1995, 1996, and 1997. These updates, however, are not comprehensive.

18. A technical definition of T is provided in the appendix at the end of Chapter 3.

19. Note that data on the residue management factor in the NRI are not disaggregated by tillage subcategories. Thus, the value of C is reported only for general tillage categories such as conventional tillage and conservation tillage and not, for example, conventional tillage with moldboard plow and no tillage.

20. Ribaudo and Hellerstein (1992) discuss the methodological issues involved in estimating the water quality benefits associated with reduced sheet and rill erosion.

21. A number of specific types of soil benefit from conventional tillage. These include soils that are generally wet in the spring and that can benefit from the creation of elevated seedbeds in the fall; soil on which the crops produce insufficient residue to control wind erosion so that tillage is required to bring up clods; soils near residential areas on which manure is disposed (this avoids the objectionable odor); soils on which residues on the surface cool the soil or carry over diseases or insects that significantly decrease crop production; soils on which high-value crops with small seeds need precise seeding depths and cannot tolerate clumps of residues; soils that become very compacted when the moisture content declines; and soils on which weeds or trees cannot be killed by herbicides (Kemper, 1995).

22. Conservation compliance is discussed more extensively in the next chapter. Conservation tillage is a significant factor in the conservation management systems and technical practices used for compliance.

23. The changes in the rate of soil erosion are computed using the universal soil loss equation.

24. A few species, such as the horned lark and the killdeer, are adversely affected by tillage practices that leave more residue on the soil. These species prefer open habitats (Best, 1995).

25. For example, there is no estimate of the bird-watching demand for wildlife in a neoclassical microeconomic setting.

26. Some studies do exist. Reicosky and colleagues (1995) and Reicosky (1995) have surveys of these.

27. One study that does focus on tillage methods is Reicosky and Lindstrom (1993). It indicates major gaseous loss of carbon immediately following tillage. The study reports the results of the effects of fall tillage methods on carbon dioxide flux from a Hamerly clay loam in the Northern Corn Belt, comparing moldboard plow, moldboard plow plus disk harrow, disk harrow only, and chisel plow with an area not tilled. Measurements immediately after tillage and intermittently for nineteen days showed that differences in carbon dioxide losses were related to soil fracturing that facilitated the movement of carbon dioxide out of an oxygen into the soil. The moldboard plow treatment buried nearly all of the residue and left the soil in a rough, loose, and open condition and resulted in the maximum carbon dioxide loss. The amounts released during those nineteen days can be compared with the equivalent carbon loss in tops and roots of the previous wheat crop (Reicosky et al., 1995). With plowing only, the carbon dioxide loss was greater than the equivalent carbon input from the previous crop. The carbon released as carbon dioxide during the nineteen days following moldboard plow, moldboard plow plus disk harrow, chisel plow, and not-tilled treatments would account for 134, 70, 58, 54, and 27 percent, respectively, of the carbon in the current year's crop residue.

28. Kern and Johnson use a number of disparate studies to estimate the change in soil organic carbon. The studies relied upon come from a number of different countries using different production technologies. Also, very different soil types and crops are aggregated together. They find that converting about 70 percent of

cropland from conventional tillage to conservation tillage will offset 0.7 percent to 1.1 percent of the U.S. fossil fuel emissions in the United States between 1990 and 2020.

29. From Wischmeier and Smith (1978).

30. From Skidmore and Woodruff (1980).

Chapter 4

1. Economic theory shows that the efficient solution (i.e., least cost for society to achieve a particular level of environmental quality) is when the marginal cost of pollution reduction is the same for all producers (Kneese and Bower, 1968). Each individual could have different combinations of practices and technologies. To implement such a policy, however, would have an extremely high cost for an industry as large and diverse as agriculture (Hefferman, 1984).

2. The supervision of these committees was transferred to the Farm Service Agency with the passage of the Federal Agricultural Improvement and Reform Act of 1996. The name of the committees was changed to State Technical Committees.

3. Many conservation programs to be implemented at the state and local levels require states to submit plans or project proposals and funding needs for federal approval before actual funds are transferred. For multiyear projects, annual plans of work and documentation of progress are required to receive continued funding. A summary of state programs for erosion control is provided in Magleby and colleagues (1995).

4. Authority for ACP was terminated on April 4, 1996, when its functions were subsumed by the Environmental Quality Incentives Program (EQIP).

5. Highly erodible land is cropland that has an erodibility index greater than or equal to eight. The technical definition of the erodibility index as well as other relevant terms is presented in the appendix to Chapter 3.

6. Analyses conducted using these data are described in Chapter 3 of this report.

7. Technically, the Food Security Act of 1985 is not the first instance of recognizing the off-site damages of soil erosion and, hence, the need to target conservation programs. The Soil Conservation Service in 1981 moved to target an increasing proportion of soil erosion programs to areas of high erosion rates to reduce substantial off-site damages, and the Agricultural Stabilization and Conservation Service began targeting its Agricultural Conservation Program (ACP) in 1982. The success of these efforts is assessed in Nielson (1985). Targeting became a general policy instrument, however, with the passage of the Ford Security Act of 1985.

8. Note that the sodbuster program was applicable to HEL uncultivated between 1981 and 1985.

9. In 1995, only 2.36 percent of ACP expenditures of 142.4 million went for conservation tillage practices (USDA, Farm Service Agency, 1996). Of the 84,258 farms receiving ACP payments in 1995, only 3,866 (4.6 percent) implemented a conservation tillage practice for which they were paid.

10. This is not a problem associated only with conservation tillage. Rather, it is characteristic of most agricultural programs (Council of Economic Advisors, 1997).

Summary

1. See Chapter 3 of this book for details of the study.

References

Alchien, A., "Costs and Output," in A. Alchien (Ed.), *The Allocation of Economic Resources*, Stanford University Press, Stanford, CT, 1959, pp. 12-34.

Alt, K., T. Osborn, and D. Colocicco, *Soil Erosion: What Effect on Agricultural Productivity*, Staff Report AIB-556, U.S. Department of Agriculture, Economic Research Service, Washington, DC, 1989.

Antle, J. and T. McGuckin, "Technological Innovation, Agricultural Productivity, and Environmental Quality," in G. Carlson, D. Zilberman, and J. Miranowski (Eds.), *Agricultural and Environmental Resource Economics*, Oxford University Press, Oxford, U.K., 1993, pp. 112-155.

Aw-Hassan, A. and A. Stoecker, *Long-Term Impact of Soil Erosion on Private and Social Returns from Alternative Tillage Systems in North Central Oklahoma*, Oklahoma Agricultural Experiment Station, Stillwater, OK, 1990.

Baker, D., "Overview of Rural Nonpoint Pollution in the Lake Erie Basin," in T. Logan and J. Baker (Eds.), *Effects of Conservation Tillage on Groundwater Quality*, Lewis Publishers, Chelsea, MI, 1987, pp. 12-34.

Baker, J., "Hydrologic Effects of Conservation Tillage and Their Importance to Water Quality," in T. Logan, J. Davidson, J. Baker, and M. Overcash (Eds.), *Effects of Conservation Tillage on Groundwater Quality: Nitrates and Pesticides*, Lewis Publishers, Chelsea, MI, 1987, pp. 113-124.

Baldassarre, G., R. Whyte, E. Quinlan, and E. Bolen, "Dynamics and Quality of Waste Corn Available to Postbreeding Waterfowl in Texas," *Wildlife Society Bulletin*, Vol. 11, No. 1 (1983), pp. 25-31.

Barker, J., G. Baumgardner, D. Turner, and J. Lee, "Carbon Dynamics of the Conservation and Wetland Reserve Programs," *Journal of Soil and Water Conservation*, Vol. 51, No. 11 (1996), pp. 340-346.

Barnard, C., S. Daberkow, M. Padgitt, M. Smith, and N. Uri, "Alternative Measures of Pesticide Use," *The Science of the Total Environment*, Vol. 203, No. 4 (1997), pp. 229-244.

Basore, N., L. Best, and J. Wooley, "Bird Nesting in Iowa No-tillage and Tilled Cropland," *Journal of Wildlife Management*, Vol. 50, No. 1 (1986), pp. 19-28.

Basta, N., R. Huhnke, and J. Stiegler, "Atrazine Runoff from Conservation Tillage Systems," *Journal of Soil and Water Conservation*, Vol. 52, No. 1 (1997), pp. 44-48.

Bates, J., A. Rayner and P. Custance, "Inflation and Farm Tractor Replacement in the U.S.: A Simulation Model," *American Journal of Agricultural Economics*, Vol. 61, No. 2 (1979), pp. 331-334.

Batte, M., "Technology and Its Impact on American Agriculture," in T. Smith (Ed.), *Size, Structure, and the Changing Face of American Agriculture*, Westview Press, Inc., Boulder, CO, 1993, pp. 173-201.

Batte, M., L. Forster, and K. Bacon, *Performance of Alternative Tillage Systems on Ohio Farms*, Department of Agricultural Economics and Rural Sociology, The Ohio State University, Columbus, OH, 1993.

Beattie, B., S. Thompson, and M. Boehlje, "Product Complementarity in Production: The By-product Case," *Southern Journal of Agricultural Economics*, Vol. 6, No. 2 (1974), pp. 161-165.

Beck, D., *Increasing the Efficient Utilization of Precipitation on the Great Plains and Prairies of North America*, Dakota Lakes Research Farm, Pierre, SD, 1996.

Becker, H., "WEPP: Spilling the Secrets of Water Erosion," *Agricultural Research,* Vol. 45, No. 1 (1997), pp. 4-8.

Best, L., "Impacts of Tillage Practices on Wildlife Habitat and Populations," in G. Steinhardt (Ed.), *Farming for a Better Environment,* Soil and Water Conservation Society, Ankeny, IA, 1995, pp. 53-55.

Bosch, D., Z. Cook, and K. Fuglie, "Voluntary versus Mandatory Agricultural Policies to Protect Water Quality: Adoption of Nitrogen Testing in Nebraska," *Review of Agricultural Economics*, Vol. 17, No. 1 (1995), pp. 13-24.

Bosch, D., K. Fuglie, and R. Keim, *Economic and Environmental Effects of Nitrogen Testing for Fertilizer Management*, Staff Report AGES-9413, U.S. Department of Agriculture, Economic Research Service, Washington, DC, April 1994.

Brady, S., "Important Soil Conservation Techniques That Benefit Wildlife," in *Technologies to Benefit Agriculture and Wildlife—Workshop Proceedings*, U.S. Congress, Office of Technology Assessment, Washington, DC, 1985, pp. 97-105.

Bruce, R., G. Langdale, L. West, and W. Miller, "Surface Soil Degradation and Soil Productivity Restoration and Maintenance," *Soil Science Society of America Journal*, Vol. 59, No. 8 (1995), pp. 654-660.

Bull, L., *Residue and Tillage Systems for Field Crops*, Staff Report AGES 9310, U.S. Department of Agriculture, Economic Research Service, Washington, DC, 1993.

Bull, L. and C. Sandretto, *Crop Residue Management and Tillage Systems Trends*, Staff Report SB-930, U.S. Department of Agriculture, Economic Research Service, Washington, DC, 1996.

Bull, L., H. Delvo, C. Sandretto, and B. Lindamood, "Analysis of Pesticide Use by Tillage System in 1990, 1991, and 1992," *Agricultural Resources: Inputs Situation and Outlook*, U. S. Department of Agriculture Economic Research Service, Washington, DC, 1993.

Carter, M. and H. Kunelius, "Adapting Conservation Tillage in Cool, Humid Regions," *Journal of Soil and Water Conservation*, Vol. 45, No. 9 (1990), pp. 454-456.

Castrale, J., "Responses of Wildlife to Various Tillage Conditions," *Transactions of the North American Wildlife and Natural Resources Conference*, Vol. 50 (1985), pp. 142-156.

Caswell, M. and R. Shoemaker, *Adoption of Pest Management Strategies Under Varying Environmental Conditions*, Technical Bulletin 1827, U.S. Department of Agriculture, Economic Research Service, Washington, DC, December 1993.

Caswell, M., D. Zilberman, and G. Casterline, "The Diffusion of Resource-Quality-Augmenting Technologies," *Natural Resource Modeling*, Vol. 7, No. 3 (1993), pp. 305-329.

Center for Agricultural Business, *Purdue/Top Commercial Producer Study*, Department of Agricultural Economics, Purdue University, West Lafayette, IN, 1995.

Ciba-Geigy Corporation, *Best Management Practices to Reduce Runoff of Pesticides into Surface Water: A Review and Analysis of Supporting Research*, TR-9-92, Agricultural Group, Ciba-Geigy Corporation, Greensboro, NC, 1992.

Clark, R., J. Johnson, and J. Brundson, "Economics of Residue Management," in W. Moldenhauer and A. Black (Eds.), *Crop Residue Management to Reduce Erosion and Improve Soil Quality: Northern Great Plains*, U.S. Department of Agriculture, Agricultural Research Service, Washington, DC, 1994, pp. 27-31.

Conservation Technology Information Center, *National Crop Residue Management Survey*, West Lafayette, IN, annual surveys.

Conservation Technology Information Center, *National Crop Residue Management Survey—1996 Survey Results*, West Lafayette, IN, 1996.

Conservation Technology Information Center, *Conservation Tillage: A Checklist for U.S. Farmers*, West Lafayette, IN, 1997.

Conway, G., "Sustainability in Agricultural Development: Trade-Offs Between Productivity, Stability, and Equability," *Journal of Farming Systems Research*, Vol. 4, No. 1 (1994), pp. 1-14.

Council of Economic Advisors, *Economic Report of the President*, U.S. Government Printing Office, Washington, DC, 1997.

Crosswhite, W. and C. Sandretto, "Trends in Resource Protection Policies in Agriculture," *Agricultural Resources: Cropland, Water, and Conservation Situation and Outlook Report*, U.S. Department of Agriculture, Economic Research Service, Washington, DC, 1991.

Dickey, E., "Tillage System Definitions," in *Conservation Tillage Systems and Management*, Midwest Plan Service, Iowa State University, Ames, IA, 1992, pp. 41-46.

Dickey, E. and D. Shelton, "Targeted Education Programs to Enhance the Adoption of Conservation Practices," in J. T. Smith (Ed.), *Optimum Erosion Control at Last Cost*, American Society of Agricultural Engineers, St. Joseph, MI, 1987, pp. 214-234.

Dickey, E., P. Jasa, and D. Shelton, "Conservation Tillage Systems," in *Conservation Tillage Systems and Management*, Midwest Plan Service, Iowa State University, Ames, IA, 1992, pp. 47-53.

Doane's Agricultural Report, *Estimated Machinery Operating Costs, 1997*, Doane Agricultural Service Company, St. Louis, MO, April 1997.

Dobbs, T., J. Bischoff, L. Henning, and B. Pflueger, "Case Study of the Potential Economic and Environmental Effects of the Water Quality Incentive Program and the Integrated Crop Management Program: Preliminary Findings," Paper presented at annual meeting of the Great Plains Economics Committee, Great Plains Agricultural Council, Kansas City, MO, April 1995.

Donigian, A., T. Barnwell, R. Jackson, A. Rartwardhan, K. Weinrich, A. Rowell, R. Chinnaswamy, and C. Cole, *Assessment of Alternative Practices and Policies Affecting Soil Carbon in Agro-ecosystems in the Central United States*, Environmental Protection Agency, Washington, DC, 1994.

Dudley, L., "Learning and Productivity Changes in Metal Products," *American Economic Review*, Vol. 62, No. 7 (1972), pp. 662-689.

Dunachie, J. and W. Fletcher, "The Toxicity of Certain Herbicides to Hen's Eggs," *Annals of Applied Biology*, Vol. 66, No. 6 (1970), pp. 515-520.

Epplin, F., G. Al-Sakkat, and T. Pepper, "Impacts of Alternative Tillage Methods for Continuous Wheat on Grain Yield and Economics: Implications for Conservation Compliance," *Journal of Soil and Water Conservation*, Vol. 49, No. 4 (1994), pp. 394-399.

Esseks, J., and S. Kraft, *Opinions of Conservation Compliance Held by Producers Subject to It*, American Farmland Trust, Washington, DC, 1993.

Fawcett, R., "Overview of Pest Management Systems," in T. Logan, J. Davidson, J. Baker, and M. Overcash (Eds.), *Effects of Conservation Tillage on Groundwater Quality: Nitrates and Pesticides*, Lewis Publishers, Chelsea, MI, 1987, pp. 90-112.

Fawcett, R., B. Christensen, and D. Tierney, "The Impact of Conservation Tillage on Pesticide Runoff into Surface Water: A Review and Analysis," *Journal of Soil and Water Conservation*, Vol. 49, No. 2 (1994), pp. 126-135.

Feder, G., R. Just, and D. Zilberman, "Adoption of Agricultural Innovations in Developing Countries: A Survey," *Economic Development and Cultural Change*, Vol. 34, No. 2 (1985), pp. 255-296.

Foltz, J., J. Lee, M. Martin, and P. Preckel, "Multiattribute Assessment of Alternative Cropping Systems," *American Journal of Agricultural Economics*, Vol. 77, No. 2 (1995), pp. 408-420.

Foster, G. and S. Dabney, "Agricultural Tillage Systems: Water Erosion and Sedimentation," *Farming for a Better Environment,* Soil and Water Conservation Society, Ankeny, IA, 1995.

Fox, G., A. Weersink, G. Sarwar, S. Duff, and B. Deen, "Comparative Economics of Alternative Agriculture Production Systems: A Review," *Northeastern Journal of Agricultural and Resource Economics*, Vol. 20, No. 1 (1991), pp. 124-142.

Frye, W., "Energy Use in Conservation Tillage," in G. Steinhardt (Ed.), *Farming for a Better Environment*, Soil and Water Conservation Society, Ankeny, IA, 1995, pp. 30-34.

Gebhart, D., H. Johnson, H. Mayeux, and H. Polley, "The CRP Increases Soil Organic Carbon," *Journal of Soil and Water Conservation*, Vol. 49, No. 3 (1994), pp. 488-492.

Gilley, J., J. Doran, D. Karlen, and T. Kaspar, "Runoff, Erosion, and Soil Quality Characteristics of a Former Conservation Reserve Program Site," *Journal of Soil and Water Conservation*, Vol. 52, No. 2 (1997), pp. 189-193.

Glotfelty, D., in "The Effects of Conservation Tillage Practices on Pesticide Volatilization and Degradation," *Effects of Conservation Tillage on Ground-water Quality*, Lewis Publishers, Chelsea, MI, 1987, pp. 314-339.

Gray, R., J. Taylor, and W. Brown, "Economic Factors Contributing to the Adoption of Reduced Tillage Technologies in Central Saskatchewan," *Journal of Plant Science*, Vol. 37, No. 1 (1997), pp. 7-17.

Griffith, D. and J. Mannering, "Optimizing Fertilizer Placement," in L. P. Jones and F. Schmitz (Eds.), *Proceedings in Plant Food and Agriculture Chemical Conference*, Purdue University, West Lafayette, IN, 1974, pp. 93-97.

Griliches, Z., "Hybrid Corn: An Exploration in the Economics of Technical Change," *Econometrica*, Vol. 25, No. 4 (1957), pp. 501-522.

Halvorsen, A., "Fertilizer Management," in W. Moldenhauer and A. Black (Eds.), *Crop Residue Management to Reduce Soil Erosion and Improve Soil Quality: Northern Great Plains*, U.S. Department of Agriculture, Agricultural Research Service, Washington, DC, 1994, pp. 22-27.

Hayami, Y. and V. Ruttan, *Agricultural Development: An International Comparison*, Johns Hopkins University Press, Baltimore, MD, 1985.

Hefferman, W., "Assumptions of the Adoption/Diffusion Model and Soil Conservation," in B. English, J. Maetzold, B. Holding, and E. Heady (Eds.), *Agricultural Technology and Resource Conservation*, Iowa State University Press, Ames, IA, 1984, pp. 86-115.

Heimlich, R., "Soil Erosion and Conservation Policies in the United States," in N. Hanley (Ed.), *Farming and the Countryside: An Economic Analysis of External Costs and Benefits*, CAB International Publishers, Miami, FL, 1991, pp. 12-17.

Herndon, L., "Conservation Systems and Their Role in Sustaining America's Soil, Water, and Related Natural Resources," in *Optimum Erosion Control at Least Cost*, American Society of Agricultural Engineers, St. Joseph, MI, 1987, pp. 135-152.

Hoffman, D. and P. Albers, "Evaluation of Potential Embryotoxicity and Teratogenicity of 42 Herbicides, Insecticides, and Petroleum Contaminants to Mallard Eggs," *Environmental Contaminants and Toxicology*, Vol. 13, No. 1 (1984), pp. 15-27.

Holland, E. and D. Coleman, "Litter Placement Effects on Microbial and Organic Matter Dynamics in an Agroecosystem," *Ecology*, Vol. 68, No. 4 (1987), pp. 425-433.

Holmes, B., *Legal Authorities for Federal (USDA), State, and Local Soil and Water Conservation Activities: Background for the Second RCA Appraisal*, U.S. Department of Agriculture, Washington, DC, 1987.

Hudson, E. and J. Bradley, "Economics of Surface-Residue Management," in G. Langdale and W. Moldenhauer (Eds.), *Crop Residue Management to Reduce Erosion and Improve Soil Quality: Southeast*, U.S. Department of Agriculture, Agricultural Research Service, Washington, DC, 1995, pp. 11-12.

Hunt, D., "Farm Machinery Technology: Performance in the Past, Promise for the Future," in B. English (Ed.), *Future Agricultural Technology and Resource Conservation*, Iowa State University Press, Ames, IA, 1984, pp. 214-229.

Hunt, P., D. Karlen, T. Matheny, and V. Quisenberry, "Changes in Carbon Content of a Norfolk Loamy Sand after 14 years of Conservation or Conventional Tillage," *Journal of Soil and Water Conservation*, Vol. 51, No. 2 (1996), pp. 255-258.

Huzsar, P. and S. Piper, "Estimating Offsite Costs of Wind Erosion in New Mexico," *Journal of Soil and Water Conservation*, Vol. 41, No. 3 (1986), pp. 414-416.

Hyberg, B. and P. Johnston, "Conservation Compliance," in *Agricultural Resources and Environmental Indicators*, U.S. Department of Agriculture, Economic Research Service, Washington, DC, 1997, pp. 297-309.

Iowa State University, *Southeastern Iowa Conservation Tillage Research Project Middletown, Iowa*, Agriculture and Home Economics Experiment Station, Ames, IA, annual reports.

Ismail, I., R. Blevins, and W. Frye, "Long Term No-Tillage Effects on Soil Properties and Continuous Corn Yields," *Soil Science Society of America Journal*, Vol. 58, No. 2 (1994), pp. 193-198.

John Deere and Company, *Fundamentals of Machinery Operation-Machine Management*, Moline, IL, 1980.

Just, R. and D. Zilberman, "Stochastic Structure, Farm Size and Technology Adoption in Developing Agriculture," *Oxford Economic Papers*, Vol. 35, No. 2 (1983), pp. 307-328.

Karlen, D., "Conservation Tillage Research Needs," *Journal of Soil and Water Conservation*, Vol. 45, No. 3 (1990), pp. 365-369.

Kemper, W., "A Bottom-Line Assessment of Advantages and Disadvantages of Various Tillage Systems," in G. Steinhardt (Ed.), *Farming for a Better Environment*, Soil and Water Conservation Society, Ankeny, IA, 1995, pp. 61-64.

Kern, J. and M. Johnson, "Conservation Tillage Impacts on National Soil and Atmospheric Carbon Levels," *Soil Science Society of America Journal*, Vol. 57, No. 2 (1993), pp. 200-221.

Kislev, Y. and N. Schori-Barach, "The Process of an Innovation Cycle," *American Journal of Agricultural Economics*, Vol. 55, No. 1 (1973), pp. 28-37.

Kneese, A. and B. Bower, "Standards, Charges, and Equity," *Managing Water Quality: Economics, Technology, Institutions,* Johns Hopkins University Press, Baltimore, MD, 1968.

Lal, R., T. Logan, and N. Fausey, "Long-Term Tillage Effects on a Mollic Ochraqualf in Northwest Ohio," *Soil and Tillage Research*, Vol. 15, No. 4 (1990), pp. 371-382.

Langdale, G., L. West, R. Bruce, W. Miller, and A. Thomas, "Restoration of Eroded Soil with Conservation Tillage," *Soil Technology*, Vol. 5, No. 1 (1992), pp. 81-90.

Libby, L., "Interaction of RCA with State and Local Conservation Programs," in H. Halcrow, E. Heady, and M. Cotner (Eds.), *Soil Conservation Policies, Institutions, and Incentives*, Soil Conservation Society of America, Ankeny, IA, 1982, pp. 119-133.

Lindstrom, M., T. Schumacher, A. Jones, and C. Gantzer, "Productivity Index Model for Selected Soils in North Central United States," *Journal of Soil and Water Conservation*, Vol. 47, No. 3 (1992), pp. 491-494.

Logan, T., "Agricultural Best Management Practices and Groundwater Protection," *Journal of Soil and Water Conservation*, Vol. 45, No. 2 (1990), pp. 201-206.

Magleby, R., C. Sandretto, W. Crosswhite, and T. Osborn, *Soil Erosion and Conservation in the United States*, Staff Report AIB-718, U.S. Department of Agriculture, Economic Research Service, Washington, DC, 1995.

Mannering, J., D. Schertz, and B. Julian, "Overview of Conservation Tillage," in *Effects of Conservation Tillage on Groundwater Quality*, Lewis Publishers, Chelsea, MI, 1987, pp. 21-39.

Martin, A., "Weed Control," in *Conservation Tillage Systems and Management*, Midwest Plan Service, Iowa State University, Ames, IA 1992, pp. 57-66.

Martin, A., H. Zim, and A. Nelson, *American Wildlife and Plants*, McGraw-Hill Book Company, New York, 1951.

McConnell, K., "An Economic Model of Soil Conservation," *American Journal of Agricultural Economics*, Vol. 65, No. 1 (1983), pp. 83-89.

Mengel, D., J. Moncrief, and E. Schulte, "Fertilizer Management," *Conservation Tillage Systems and Management*, Midwest Plan Service, Iowa State University, Ames, IA, 1992, pp. 83-87.

Midwest Plan Service, *Conservation Tillage Systems and Management*, Iowa State University, Ames, IA, 1992.

Mikesell, C., J. Williams, and J. Long, "Evaluation of Net Return Distributions from Alternative Tillage Systems for Grain Sorghum and Soybean Rotations," *North Central Journal of Agricultural Economics*, Vol. 10, No. 2 (1988), pp. 255-271.

Miranowski, J., "Economics of Land in Agriculture," in G. Carlson, D. Zilberman, and J. Miranowski (Eds.), *Agricultural and Environmental Resource Economics*, Oxford University Press, Oxford, U.K., 1993, pp. 295-341.

Missouri Management Systems Evaluation Area (MSEA), *A Farming Systems Water Quality Report*, Research and Education Report, Missouri MSEA, 1995.

Moldenhauer, W. and A. Black, *Crop Residue Management to Reduce Erosion and Improve Soil Quality: Northern Great Plains*, U.S. Department of Agriculture, Agricultural Research Service, Washington, DC, 1994.

Moldenhauer, W., W. Kemper, and G. Langdale, "Long Term Effects of Tillage and Crop Residue Management," in G. Langdale and W. Moldenhauer (Eds.), *Crop Residue Management to Reduce Erosion and Improve Soil Quality—Southeast*, Staff Report, CR-39, U.S. Department of Agriculture, Agricultural Research Service, Washington, DC, 1994, pp. 38-48.

Monson, M. and N. Wollenhaupt, "Residue Management: Does It Pay?" *Crop Residue Management to Reduce Soil Erosion and Improve Soil Quality: North*

Central Region, U.S. Department of Agriculture, Agricultural Research Service, Washington, DC, 1995.

National Research Council, *Soil and Water Quality,* National Academy Press, Washington, DC, 1993.

Nelson, F. and L. Schertz, *Provisions of the Federal Agriculture Improvement and Reform Act of 1996,* U.S. Department of Agriculture, Economic Research Service, Washington, DC, 1996.

Nielson, J., *Targeting Erosion Control: Delivering Technical and Financial Assistance to Farmers,* U.S. Department of Agriculture, Agricultural Research Service, Washington, DC, 1985.

Nowak, P., "Adoption and Diffusion of Soil and Water Conservation Practices," in B. English (Ed.), *Agricultural Technology and Resource Conservation,* Iowa State University Press, Ames, IA, 1984, pp. 51-59.

Nowak, P., "Why Farmers Adopt Production Technology," *Journal of Soil and Water Conservation,* Vol. 47, No. 1 (1992), pp. 14-16.

Nowak, P. and G. O'Keefe, "Evaluation of Producer Involvement in the United States Department of Agriculture 1990 Water Quality Demonstration Projects," Baseline Report submitted to the U.S. Department of Agriculture, Washington, DC, November 1992.

Nowak, P. and G. O'Keefe, "Farmers and Water Quality: Local Answers to Local Issues," Draft Report submitted to the U.S. Department of Agriculture, Washington, DC, September 1995.

Odell, R., W. Walker, L. Boone, and M. Oldham, *The Morrow Plots: A Century of Learning,* Agricultural Experiment Station Bulletin, University of Illinois, Urbana, IL, 1984.

Office of Technology Assessment, *Beneath the Bottom Line: Agricultural Approaches to Reduce Agrichemical Contamination of Groundwater,* OTA-F-418, U.S. Congress, Washington, DC, November 1990.

Olson, K. and N. Senjem, *Economic Comparison of Incremental Changes in Tillage Systems in the Minnesota River Basin,* Minnesota Extension Service, University of Minnesota, Minneapolis, 1996.

Osborn, T., "Conservation," in *Provisions of the Federal Agricultural Improvement Act of 1996,* U.S. Department of Agriculture, Economic Research Service, Washington, DC, 1996, pp. 41-46.

Pagoulatos, A., D. Debertin, and F. Sjarkowi, "Soil Erosion, Intertemporal Profit, and the Soil Conservation Decision," *Southern Journal of Agricultural Economics,* Vol. 21, No. 1 (1989), pp. 55-62.

Paudel, K. and L. Lohr, *Economic Effects of Imperfect Information on Conservation Decisions,* Department of Agricultural and Applied Economics, University of Georgia, Athens, GA, 1996.

Pike, D., W. Kirby, and S. Kamble, *Distribution and Severity of Pests in the Midwest,* College of Agricultural, Consumer and Environmental Sciences, University of Illinois, Urbana, 1997.

Piper, S., "Enrolling Land in the CRP Provides Economic Benefits when Wind Erosion Is Reduced," in M. Ribaudo, D. Colacicco, L. Langner, S. Piper, and

G. Schaible (Eds.), *Natural Resources and Users Benefits from the Conservation Reserve Program,* Staff Report AER-627, U.S. Department of Agriculture, Economic Research Service, Washington, DC, 1990, pp. 35-51.

Piper, S. and L. Lee, *Estimating the Offsite Household Damage from Wind Erosion in the Western United States,* Staff Report 8926, U.S. Department of Agriculture, Economic Research Service, Washington, DC, 1989.

Raich, J. and C. Potter, *Global Patterns of Carbon Dioxide Emissions from Soils on a 0.5 Degree Grid Cell Basis,* Oak Ridge National Laboratory, Oak Ridge TN, 1997.

Rasiah, R. and B. Kay, "Characterizing the Changes in Aggregate Stability Subsequent to Introduction of Forages," *Soil Science Society of America Journal,* Vol. 58, No. 10 (1994), pp. 935-942.

Rasmussen, W., "History of Soil Conservation, Institutions and Incentives," in T. Lutz (Ed.), *Soil Conservation Policies, Institutions, and Incentives,* Soil Conservation Society of America, Ankeny, IA, 1982, pp. 4-7.

Rasmussen, P., R. Allmaras, C. Rhode, and N. Roager, "Crop Residue Influences on Soil Carbon and Nitrogen in a Wheat-Fallow System," *Soil Science Society of America Journal,* Vol. 44, No. 6 (1980), pp. 596-600.

Rasmussen, P. and H. Collins, "Long-Term Impacts of Tillage, Fertilizer and Crop Residue on Soil Organic Matter in Temperate Semi-Arid Regions," *Advances in Agronomy,* Vol. 45, No. 2 (1991), pp. 93-134.

Rehm, G., "Tillage Systems and Fertilizer Management," in G. Steinhardt (Ed.), *Farming for a Better Environment,* Soil and Water Conservation Society, Ankeny, IA, 1995, pp. 23-24.

Reichelderfer, K., "Land Stewards or Polluters? The Treatment of Farmers in the Evaluation of Environmental and Agricultural Policy," Paper presented at the conference Is Environmental Quality Good for Business?, Washington, DC, 1990.

Reicosky, D., "Impact of Tillage on Soil As a Carbon Sink," in G. Steinhardt (Ed.), *Farming for a Better Environment,* Soil and Water Conservation Society of America, Ankeny, IA, 1995, pp. 50-52.

Reicosky, D., W. Kemper, G. Langdale, C. Douglas, and P. Rasmussen, "Soil Organic Matter Changes Resulting from Tillage," *Journal of Soil and Water Conservation,* Vol. 50, No. 2 (1995), pp. 253-261.

Reicosky, D. and M. Lindstrom, "The Effect of Fall Tillage Methods on Short-Term Carbon Dioxide Flux from Soil," *Agronomy Journal,* Vol. 85, No. 11 (1993), pp. 1237-1243.

Renard, K., G. Foster, G. Weesies, D. McCool, and D. Yoder, *Predicting Soil Erosion by Water: A Guide to Conservation Planning with the Revised Universal Soil Loss Equation (RUSLE),* Agriculture Handbook 703, U.S. Department of Agriculture, Washington, DC, 1996.

Ribaudo, M., *Water Quality Benefits from the Conservation Reserve Program,* Staff Report AER-606, U.S. Department of Agriculture, Economic Research Service, Washington, DC, 1989.

Ribaudo, M., D. Colacicco, L. Langner, S. Piper, and G. Schaible, *Natural Resources and Users Benefit from the Conservation Reserve Program*, Staff Report AER-627, U.S. Department of Agriculture, Economic Research Service, Washington, DC, 1990.

Ribaudo, M. and D. Hellerstein, *Estimating Water Quality Benefits: Theoretical and Methodological Issues*, Staff Report TB-1808, U.S. Department of Agriculture, Economic Research Service, Washington, DC, 1992.

Richards, R. and D. Baker, *Twenty Years of Change: The Lake Erie Agricultural Systems for Environmental Change (LEASEQ) Project*, Water Quality Laboratory, Heidelberg College, Tiffin, OH, 1998.

Roberts, W. and S. Swinton, "Economic Methods for Comparing Alternative Crop Production Systems," *American Journal of Alternative Agriculture*, Vol. 11 (1996), pp. 10-17.

Rodgers, R. and J. Wooley, "Conservation Tillage Impacts on Wildlife," *Journal of Soil and Water Conservation*, Vol. 38, No. 2 (1983), pp. 212-213.

Rosen, S., "Learning by Experience," *Quarterly Journal of Economics*, Vol. 86, No. 3 (1972), pp. 366-382.

Saha, A., H.A. Love, and R. Schwart, "Adoption of Emerging Technologies Under Output Uncertainty," *American Journal of Agricultural Economics*, Vol. 76, No. 5 (1994), pp. 836-846.

Schepers, J., "Effects of Conservation Tillage on Processes Affecting Nitrogen Management," *Effects of Conservation Tillage on Groundwater Quality*, Lewis Publishers, Chelsea, MI, 1987.

Schertz, D., "Conservation Tillage: An Analysis of Acreage Projections in the United States," *Journal of Soil and Water Conservation*, Vol. 33, No. 2 (1988), pp. 256-258.

Schumacher, T., M. Lindstrom, M. Blecha, and R. Blevins, "Management Options for Lands Concluding Their Tenure in the Conservation Reserve Program," in R. Blevins and W. Moldenhauer (Eds.), *Crop Residue Management to Reduce Erosion and Improve Soil Quality—Appalachia and Northeast*, U.S. Department of Agriculture, Agricultural Research Service, Washington, DC, 1995, pp. 7-10.

Setia, P., "Evaluating Soil Conservation Management Systems Under Uncertainty," in *Optimum Erosion Control and Least Cost*, American Society of Agricultural Engineers, St. Joseph, MI, 1987, pp. 402-422.

Siemans, J., "Soil Management and Tillage Systems," *Illinois Agronomy Handbook*, University of Illinois Press, Urbana, 1997

Siemans, J. and D. Doster, "Costs and Returns," in *Conservation Tillage Systems and Management*, Midwest Plan Service, Iowa State University, Ames, IA, 1992, pp. 30–52.

Skidmore, E. and N. Woodruff, *Wind Erosion Forces in the United States and Their Use in Predicting Soil Loss*, U.S. Department of Agriculture Handbook 346, U.S. Department of Agriculture, Agricultural Research Service, Washington, DC, 1980.

Smith, E. and B. English, *Determining Wind Erosion in the Great Plains,* Card Paper 82-3, Center for Agriculture and Rural Development, Iowa State University, Ames, IA, 1982.

Snider, R., J. Moore, and J. Subagja, "Effects of Paraquat and Atrazine on Nontarget Soil Arthropods," in F. D'Itri (Ed.), A Systems Approach to Conservation Tillage, Lewis Publishers, Chelsea, MI, 1985, pp. 191-196.

State of Illinois, *Carbon Emissions from Fossil Fuel Consumption,* Department of Natural Resources, State of Illinois, Springfield, IL, 1997.

Strohbehn, R., *An Economic Analysis of USDA Erosion Control Programs: A New Perspective,* Staff Report AER-560, U.S. Department of Agriculture, Economic Research Service, Washington, DC, 1986.

Tate, R., *Soil Organic Matter: Biological and Ecological Effects,* John Wiley and Sons, New York, 1987.

Tracy, M., *Food and Agriculture in a Market Economy,* Agricultural Policy Studies, La Hutte, Belgium, 1993.

Triazine Network News, "Economic, Environmental Issues Raised in Recent Special Review Submission," Chicago, IL, December 1996.

Tweeten, L., "The Structure of Agriculture: Implications for Soil and Water Conservation," Journal of Soil and Water Conservation, Vol. 50, No. 4 (1995), pp. 347-351.

U.S. Department of Agriculture, Economic Research Service, *Cropping Practices Survey: Data File Specification and Documentation,* Washington, DC, annual surveys.

U.S. Department of Agriculture, Economic Research Service, *Agricultural Resources and Environmental Indicators,* Washington, DC, 1997.

U.S. Department of Agriculture, Farm Service Agency, *Agricultural Conservation Program Statistical Summary Report,* Washington, DC, 1996.

U.S. Department of Agriculture, Natural Resources Conservation Service, *The 1992 National Resources Inventory,* published data, Washington, DC, 1994.

U.S. Department of Agriculture, Natural Resources Conservation Service, *1995 Status Review Results,* Washington, DC, 1996.

U.S. Department of Agriculture, Natural Resources Conservation Service, "Residue Management," *Conservation Practice Standards,* Washington, DC, 1997a, pp. 14-16.

U.S. Department of Agriculture, Natural Resources Conservation Service, *A Geography of Hope,* U.S. Washington, DC, 1997b.

U.S. Department of Agriculture, "No-Till's Benefits and Costs," *The USDA Resource Conservation Systems Application Program,* Washington, DC, 1997, pp. 11-17.

U.S. Department of Agriculture, Soil Conservation Service, *Instructions for Collecting 1992 National Resources Inventory Sample Data,* Washington, DC, 1992.

U.S. Department of the Interior, Fish and Wildlife Service, *National Survey of Fishing, Hunting, and Wildlife-Associated Recreation—1991,* Washington, DC, 1993.

U.S. Environmental Protection Agency, *National Water Quality Inventory—1994 Report to Congress,* Washington, DC, 1995.

U.S. Environmental Protection Agency, *Notes on Agriculture,* Washington, DC, 1996.

University of Illinois Agricultural Extension Service, *No-Till's Benefits and Costs,* Urbana, IL, 1997.

Uri, N., "Conservation Tillage and Input Use," *Environmental Geology,* Vol. 29, No. 3 (1997), pp. 188-200.

Uri, N.D. and K. Day, "Energy Efficiency, Technological Change and the Dieselization of American Agriculture in the United States," *Transportation Planning and Technology,* Vol. 16, No. 2 (1992), pp. 221-231.

Uri, N. and B. Hyberg, "Stream Sediment Loading and Rainfall—A Look at the Issue," *Water, Air, and Soil Pollution,* Vol. 51, No. 2 (1990), pp. 95-104.

van Es, J. "Dilemmas in the Soil and Water Conservation Behavior of Farmers," in B. English, J. Maetzold, B. Holding, and E. Heady (Eds.), *Agricultural Technology and Resource Conservation,* Iowa State University Press, Ames, IA 1984, pp. 61-93.

Veseth, R., B. Miller, T. Fiez, and T. Walters, "Returning CRP Land to Crop Production," *Pacific Northwest Conservation Tillage Handbook,* Washington State University, Pullman, WA, 1997.

Waddington, D., K. Boyle, and J. Cooper, *1991 Net Economic Values for Bass and Trout Fishing, Deer Hunting, and Wildlife Watching,* U.S. Department of the Interior, Fish and Wildlife Service, Washington, DC, 1994.

Warburton, D., and W. Klimstra, "Wildlife Use of No-Till and Conventionally Tilled Corn Fields," *Journal of Soil and Water Conservation,* Vol. 39, No. 4 (1984), pp. 327-330.

Washington State University Agriculture Extension Service, *Pacific Northwest Conservation Tillage Handbook,* Washington State University, Pullman, WA, 1997.

Wauchope, R.D., "Effects of Conservation Tillage on Pesticide Loss with Water," in T. Logan, J. Davidson, J. Baker, and M. Overcash (Eds.), *Effects of Conservation Tillage on Groundwater Quality: Nitrates and Pesticides,* Lewis Publishers, Chelsea, MI, 1987, pp. 201-215.

Weersink, A., M. Walker, C. Swanton, and F. Shaw, "Costs of Conventional and Conservation Tillage Systems," *Journal of Soil and Water Conservation,* Vol. 47, No. 2 (1992), pp. 145-152.

Westra, J. and K. Olson, *Farmers' Decision Processes and the Adoption of Conservation Tillage,* Department of Applied Economics, University of Minnesota, Minneapolis, MN, 1997.

Williams, J., R. Llewelyn, L. Goss, and J. Long, *Analysis of Net Returns to Conservation Tillage from Corn and Soybeans in Northeast Kansas,* Kansas Agricultural Experiment Station Bulletin, Kansas State University, Manhattan, KS, 1988.

Williams, J. and C. Mikesell, "Conservation Tillage in the Great Plains," in *Optimum Erosion Control and Least Cost,* American Society of Agricultural Engineers, St. Joseph, MI, 1987, pp. 175-192.

Wischmeier, W., and D. Smith, *Predicting Rainfall Erosion Losses,* Agriculture Handbook 537, U.S. Department of Agriculture, Agricultural Research Service, Washington, DC, 1978.

Wittmuss, H., "Erosion Control Incorporating Conservation Tillage, Crop Rotation and Structural Practices," *Optimum Erosion Control at Least Cost,* American Society of Agricultural Engineers, St. Joseph, MI, 1987, pp. 193-219.

Wolf, S., "Cropping Systems and Conservation Policy: The Roles of Agrichemical Dealers and Crop Consultants," *Journal of Soil and Water Conservation,* Vol. 50, No. 2 (1995), pp. 263-270.

Yoder, D. and J. Lown, "The Future of RUSLE: Inside the New Revised Universal Soil Loss Equation," *Journal of Soil and Water Conservation,* Vol. 50, No. 4 (1995), pp. 484-489.

Young, D., T. Kwon, and F. Young, "Profit and Risk for Integrated Conservation Farming Systems in the Palouse," *Journal of Soil and Water Conservation,* Vol. 49, No. 2 (1994), pp. 165-168.

Young, D., D. Wallace, and P. Kanjo, "Cost Effectiveness and Equity Aspects of Soil Conservation Programs in a Highly Erodible Region," *American Journal of Agricultural Economics,* Vol. 73, No. 6 (1991), pp. 784-793.

Young, R., "Response of Small Mammals to No-Till Agriculture in Southwestern Iowa," MS Thesis, Iowa State University, Ames, 1984.

Young, R. and W. Clark, "Rodent Populations and Crop Damage in Minimum Tillage Cornfields," Paper presented to the Midwest Fish and Wildlife Conference, St. Louis, MO, 1983.

Zero Tillage Farmers Association, *Advancing the Art,* Author, Bismark, ND, 1997.

Zison, S., K. Haven, and W. Mills, *Water Quality Assessment: A Screening Method for Nondesignated 208 Areas,* Environmental Protection Agency, Athens, GA, 1977.

Zobeck, T., N. Rolong, D. Fryrear, J. Bilbro, and B. Allen, "Properties of Recently Tilled Sod, 70-Year Cultivated Soil," *Journal of Soil and Water Conservation,* Vol. 50, No. 2 (1995), pp. 210-215.

Index

Page numbers followed by the letter "n" indicate notes.

Florida, 13, 14
Food Security Act of 1985,
    70-71, 88-89, 99-100
Forage, 11, 12
Forster, L., 23
Foster, G., 56
Fox, G., 32, 94
Fruit, 61
Frye, W., 29, 30, 76
Fuel, 25, 48, 76
Fuglie, K., 38, 82
Fungal decomposition, 75

Gebhart, D., 87
Gilley, J., 76
Glotfelty, D., 57
Goss, L., 17, 25
Grain, 11, 17, 24
Grass, 87. *See also* Pasture
Gray, R., 25
Griffith, D., 38
Griliches, Z., 23

Halvorsen, A., 38-39
Haven, K., 59
Hayami, Y., 28
Hefferman, W., 86
Heimlich, R., 88
Herbicides
    as cost factor, 25, 48-50, 51
    and environment, 58, 76
    with tillage types, 41, 100
Herndon, L., 57
Hoffman, D., 73
Holland, E., 75
Holmes, B., 86
Hunt, D., 48
Hunt, P., 75
Huydson, E., 30
Huzsar, P., 2, 65, 68
Hyberg, B., 17

Illinois, 13, 14, 30, 76
Indiana, 13, 14, 18
Information sources, 8-9, 95-96
Insecticides, 41
Iowa, 13, 14, 89-93
Ismail, I., 29, 30

Jasa, P., 36
John Deere and Co., 48
Johnson, J., 32
Johnson, M., 76
Johnston, P., 17
Julian, B., 6
Just, R., 23, 24

Kamble, S., 42
Kanjo, P., 97
Kansas, 13, 14, 25
Karlen, D., 88
Kay, B., 87
Keim, R., 38
Kemper, W., 3, 88
Kentucky, 13, 14, 24, 100
Kern, J., 76
Kirby, W., 42
Kislev, Y., 23
Klimstra, W., 71
Kraft, S., 17
Kunelius, H., 25
Kwon, T., 32

Labor, 36-38, 51
Lake states, 15, 17
Lal, R., 30
Land retirement, 84
Langdale, G., 3, 6, 29, 88
Lee, L., 58
Libby, L., 86
Lindstrom, M., 87
Llewelyn, R., 17, 25
Logan, T., 30, 95
Lohr, L., 29

Order Your Own Copy of
This Important Book for Your Personal Library!

CONSERVATION TILLAGE IN U.S. AGRICULTURE
Environmental, Economic, and Policy Issues

_____ in hardbound at $89.95 (ISBN: 1-56022-884-9)

_____ in softbound at $39.95 (ISBN: 1-56022-897-0)

COST OF BOOKS_____

OUTSIDE USA/CANADA/
MEXICO: ADD 20%_____

POSTAGE & HANDLING_____
*(US: $3.00 for first book & $1.25
for each additional book)
Outside US: $4.75 for first book
& $1.75 for each additional book)*

SUBTOTAL_____

IN CANADA: ADD 7% GST_____

STATE TAX_____
*(NY, OH & MN residents, please
add appropriate local sales tax)*

FINAL TOTAL_____
*(If paying in Canadian funds,
convert using the current
exchange rate. UNESCO
coupons welcome.)*

☐ **BILL ME LATER:** ($5 service charge will be added)
(Bill-me option is good on US/Canada/Mexico orders only;
not good to jobbers, wholesalers, or subscription agencies.)

☐ Check here if billing address is different from
shipping address and attach purchase order and
billing address information.

Signature_____

☐ **PAYMENT ENCLOSED: $**_____

☐ **PLEASE CHARGE TO MY CREDIT CARD.**

☐ Visa ☐ MasterCard ☐ AmEx ☐ Discover
☐ Diners Club
Account #_____

Exp. Date_____

Signature_____

Prices in US dollars and subject to change without notice.

NAME _____

INSTITUTION _____

ADDRESS _____

CITY _____

STATE/ZIP _____

COUNTRY _____ COUNTY (NY residents only) _____

TEL _____ FAX _____

E-MAIL_____

May we use your e-mail address for confirmations and other types of information? ☐ Yes ☐ No

Order From Your Local Bookstore or Directly From
The Haworth Press, Inc.
10 Alice Street, Binghamton, New York 13904-1580 • USA
TELEPHONE: 1-800-HAWORTH (1-800-429-6784) / Outside US/Canada: (607) 722-5857
FAX: 1-800-895-0582 / Outside US/Canada: (607) 772-6362
E-mail: getinfo@haworthpressinc.com
PLEASE PHOTOCOPY THIS FORM FOR YOUR PERSONAL USE.

BOF96